PEOPLE IN SPACE

PEOPLE IN SPACE

Policy Perspectives for a "Star Wars" Century

Edited by

James Everett Katz

Foreword by Hans Mark

Transaction Books
New Brunswick (U.S.A.) and Oxford (U.K.)

To my beloved Laura

Copyright © 1985 by Transaction, Inc.
New Brunswick, New Jersey 08903

Library of Congress Catalog Number: 85-1011
ISBN: 0-88738-052-2 (cloth), 0-88738-609-1 (paper)
Printed in the United States of America

Library of Congress Cataloging in Publication Data
Main entry under title:

People in space.

1. Astronautics, Military—Addresses, essays, lectures. 2. Astronautics—Government policy—Addresses, essays, lectures. I. Katz, James Everett.
UG1520.P46   1985   358'.8   85–1011
ISBN 0–88738–052––2
ISBN 0–88738–609–1 (pbk.)

Contents

Foreword vii
Hans Mark
Preface 1
James Everett Katz

Introduction 7
1. American Space Policy at the Crossroads........................ 9
James Everett Katz

Part I   Military Activities in Space 33
2. Space and the Preservation of Freedom.......................... 35
Harrison H. Schmitt
3. Space Militarization: A Costly Mistake 40
John Joseph Moakley
4. Space Militarization and the Maintenance of Deterrence.......... 44
Thomas Blau and *Daniel Gouré*
5. Space Militarization and International Law 55
Harry H. Almond, Jr.

Part II   Civilian Activities in Space 65
6. The Space Station: Mankind's Permanent Presence in Space 67
Hans Mark
7. U.S. International Space Activities 82
Marcia S. Smith
8. The Proliferation of Communications Satellites: Gold Rush in the
   Clarke Orbit 98
Joseph N. Pelton
9. Underdevelopment via Satellite: The Interests of the German
   Space Industry in Developing Countries and Their Consequences   110
Jürgen Häusler and *Georg Simonis*
10. The Controversy over Remote Sensing 129
Jean-Louis Magdelénat

11. The Moon Treaty: Reflections on the Proposed Moon Treaty,
Space Law, and the Future 140
Nathan C. Goldman

Part III   The Sociology of Outer Space 151
12. Beyond Bureaucratic Policy: The Space Flight Movement........ 153
William Sims Bainbridge
13. The Social Forces behind Technological Change and Space
Policymaking 164
James Everett Katz
14. Extraterrestrial Intelligence: The Social Impact of an Idea........ 178
Ron Westrum, David Swift, and *David Stupple*
15. The Social Psychology of Space Travel.......................... 194
B. J. Bluth
16. Must There Be Space "Colonies"? A Jurisprudential Drift to
Historicism 207
George S. Robinson

About the Contributors 222

Foreword

Hans Mark

It is always difficult to discern the major turning points in human history while they are happening. Was Thomas Newcomen aware of the implications when he used steam to drive a pump to remove water from Cornish tin mines during the early years of the eighteenth century? Did he realize that 100 years later the first commercial railroad using steam engines would be put in service? Would he have believed that only 50 years after that, millions of people would be traveling on steam-driven trains and that his invention would have by then ushered in the industrial revolution of the nineteenth century? I doubt it. Even the Wright brothers, who were somewhat more aware of the consequences of what they were doing than their distinguished inventive predecessor, did not dream that three-quarters of a century after they made their first flight in a primitive aircraft millions of people would routinely use aircraft to get around.

We started sending people into space 24 years ago. At the present time, about 200 humans have experienced space flight. How many people will have experienced space flight by the year 2011, 50 years after mankind's first venture into space? If we can believe NASA's current projections, the Shuttle will fly about 24 times a year and the average crew per flight will be about five people. Thus somewhat over 100 people per year will be flying in space which means that in the coming 25 years, something like 2,500 to 3,000 Americans will experience space flight. Double that to include the Russians and you get 5,000 to 6,000 people.

And, I think this estimate is almost surely conservative. It assumes, for example, that the American military will stick to its current position that people in space are not necessary to carry out missions related to the national defense. I have also assumed that NASA's projected flight rate—which is a constant number of flights per year—is correct. This obviously leads to a linear growth in the number of people going into space which is clearly at variance with the historical experience for enterprises such as this. In almost all cases (say the introduction of air transportation, or the introduction of automobiles), the initial growth was exponential, not linear. What happens if we make an estimate based on the assumption of exponential growth of space flight given the data points we have today for the number of people that will have been in space by the year 2011? A rough estimate yields about 50,000 people. In my judgment, this number is much closer to what we will see than my earlier one. In fact, I am willing to bet on it!

While it is fun to play with numbers, it is, in the end, not very instructive. This is a book about people in space. The important question it deals with is *why* people go into space, not *how many* will go into space. The two are, of course, related but the reasons for going into space are clearly the more important. *People In Space* provides a good and coherent picture of many of the issues involved in going into space.

I would like to focus for a moment on two reasons for space exploration that seem to me to be critical. There is much controversy today over the feasibility of constructing a space-based defense against nuclear-tipped ballistic missiles. A large number of technical experts have taken the position that such a thing cannot be done, and today, these people are probably in the majority among those who have the necessary technical background to make a sound judgment. And yet, I think this majority is wrong. I cannot prove it but I believe what we have here is an example of Arthur C. Clarke's first law of prophecy: "When a distinguished scientist says that something is possible, he is almost certainly right. When he states something is impossible, he is very probably wrong." There will come a time in the not-too-distant future when the technical pieces will fall into place and when people will wonder why anyone ever doubted the feasibility of space-based missile defense systems. My guess is that the deployment of such a system will require the presence of many thousands of people in space; this will be one of the important reasons why the exponential estimate is more accurate than the linear one.

There is another reason, however, why people will go into space, which is, to my mind, much more important, and that is the nature of human freedom. At some point, someone will discover something that can (or must) be done in space that will turn out to be enormously profitable. Space will then become the new economic frontier—rather than the intellectual or scientific and technical frontier it is today. Once this happens, an entirely new and dynamic set of rules will apply. When President Thomas Jefferson bought the Louisiana Territory in 1803, it was also an intellectual frontier rather than an economic one. A period of exploration followed but the really massive occupation of the new frontier was triggered by the discovery of gold in California in 1849 and by the search for prime agricultural land in the Oregon Territory. These things led to massive population movements that 50 years later saw the United States as a nation spanning the entire continent. Is it likely that someone will find something on the moon, say on the time scale of the next hundred years, that is so valuable that a similar population movement could result? My guess is that it is not only likely but even probable.

I submit that human freedom depends on growth and on expanding horizons. If you stifle growth, the available pie of resources becomes finite, which means that some authority has to be established to divide the pie among those who must be sustained. This is the road to totalitarianism and ultimately communism or facism. Growth means that new pies are always being created, which eliminates the necessity for a dictator and makes political freedom possible. (It is important to recognize that it does not *guarantee* freedom but makes it *possible*.) It is now widely appreciated that there are new frontiers beyond the confines of the Earth's atmosphere that will continue to beckon to the adventurous souls among us. The conquest of this new frontier will be the driving force that will continue to preserve human freedom. This, to me, is the real reason why it is important to have people in space.

Preface

It has been just two hundred years since the first human beings ascended to the heavens. In 1783, two Frenchmen sailed to the astounding height of three thousand feet over Paris in a balloon. Some people still alive today can recall the first powered flight, by the Wright brothers, in 1903. Little more than a quarter-century ago the first artificial satellite hurtled into orbit around the Earth. And less than a score of years ago the first human being set foot on another celestial object.

Today space has become a commonplace thing. Space Shuttle launchings, while always exciting, have become almost ordinary events. Communications satellites have brought every part of the globe into instantaneous communication with every other part. Observation satellites help us predict the weather better and to monitor natural phenomena on earth and throughout the rest of the universe. The uses of outer space are no longer exceptional; rather, they have become part of the quotidian life of most Americans and many people throughout the world.

Just as, historically speaking, sociology has moved from studying unusual phenomena to developing the "sociology of everyday life," so too is it time for policy analysis to move from viewing space as an exceptional phenomenon to simply a part of everyday life. This in essence is the purpose of the present volume. The authors look at the human uses of outer space as endeavors wherein ordinary policy processes and ordinary sociological processes take place. They analyze policy issues relating to outer space and give their own perspectives on how outer space policy ought to be, and is being, formed. Because space policy is a new, often emotionally charged field, it is laden with conflicting ideas, values, and world views, or one should say, galactic views. This volume gives free rein to these various and opposing views so that the reader may see the scope of issues involved and the multitude of concepts which make up that variegated and constantly evolving field known as space policy.

The book is organized into three main sections. After an overview chapter by the editor, there follow military activities in space, civilian activities in space, and space as sociological phenomena.

The overview chapter reviews space activities through a policy lens. It is noted that although space was a subject of intense interest late in the 1950s

1

and early 1960s, the study of space policy was eclipsed during the late 1960s and the 1970s. In part, this is because policy analysis has followed funding patterns. When a great deal of money is given to a policy area, usually a great deal of scholarly attention is devoted to it; likewise when funding declines, so too does scholarly interest. Of course with the rising interest in permanently manned space stations, public and scholarly interest in space is again rising.

In the overview I also show that much of space policy provides rich exploring ground for scholars: Bureaucratic politics and organizational structure analysis are models that could be heuristically applied to the space program; what happens in space has grave implications for the security of nations on Earth. Yet, until recently, little high-level attention has been given to these matters, and only under the Reagan administration has a coherent space policy taken shape. In addition, there is a broad array of implications of remote sensing (satellite detection of activities) which ranges from effects on the international economy to the exercise of individual civil rights. Other satellite-use issues include the increasing friction between developed and developing countries over the availability of geostationary satellite ''parking places'' in the Clarke orbit. Direct broadcasting from satellites may also have dramatic cultural and political consequences. In conclusion, I examine possible frameworks through which an international space policy could be developed. Because of the inadequacies of traditional arrangements in this regard, alternative approaches may be helpful. Nevertheless there is little hope of a permanent solution in the foreseeable future.

In the section on military activities in space, former U.S. Senator Harrison Schmitt lets readers share a private letter that he delivered to President Reagan in 1982. This letter helped fuel the president's interest in developing a manned permanent space station and space-based strategic defenses. Senator Schmitt, a former astronaut, is one of the few human beings ever to have walked on the moon. During his years in the Senate, he was an influential defender of space policy interests. For these reasons, his perspective on the military uses of outer space is important. He argues that space is the next strategic horizon and that the control over the space surrounding the Earth will dramatically influence the course of human events. He believes that the United States has not been vigorous enough in exercising control and expanding the technology. To effect such control, he makes a spirited argument for expansion of U.S. military space activities.

In contrast to Senator Schmitt, Congressman John Joseph Moakley urges that the United States back away from its planned deployment of antisatellite weapons and its commitment to build a space-based ballistic missile. While agreeing that additional ongoing research is necessary to prevent the Soviet Union defense from gaining a temporary advantage, he maintains that the

United States should place a moratorium on further flight tests of our anti-satellite weapons. He thinks that the United States should reinstitute space arms-control negotiations with the Soviet Union and affirm our commitment to the antiballistic missile treaty we signed with the Soviets. The attempt to enhance national security through space militarization is, says Moakley, a horrendously expensive and ultimately futile effort.

These essays are followed by an analysis by Thomas Blau and Daniel Gouré of military activities in space. They survey the importance of space control in U.S. deterrence policy and come to a mixed conclusion. They see the dramatic expansion of space capabilities advocated by the Reagan administration as premature and inherently destabilizing, but they think that long-term technologies for exercising military control over space are important and should be considered.

Although some hold forth the hope that an international organization could intervene and help resolve or head off military conflicts in space, Harry Almond casts a pessimistic light on such hopes in his essay. Almond points to the futility of many past arms control agreements, and argues that there is reason to suspect that an international organization or international agreement could help permanently prevent or resolve conflict in outer space.

Moving away from military issues, the second section of the book deals with the civilian side of space policy. Although military concerns have provided a powerful impetus to space activities, the greatest achievements have been civilian in nature, most notably the lunar landings of Project Apollo. Yet it is not always easy to distinguish between civil and military uses of space; many technologies, for example remote sensing, communications satellites, and manned space stations, can be used for either civilian or military purposes.

The genesis of the U.S. space station project, which is to be primarily civilian, is analyzed in the chapter by Hans Mark, Chancellor of the University of Texas system. Mark served as Secretary of the Air Force and for many years was a high NASA official. In his chapter he notes that although the leadership of the military and space sciences communities opposed building a manned space station, through foresight and diligence NASA's leadership was able to win President Reagan's support for a permanently manned space station. He probes in depth the historical, organizational, and personality factors that led to this great leap toward mankind's inhabitation of space.

While the space station does have some potential military applications, even purely civilian space projects such as those involving astrophysical observation or educational broadcasting can become enmeshed in diplomatic flaps and international rivalry. This can be seen in the events that overtake some U.S. international space projects, the subject of the chapter of Marcia Smith of the Congressional Research Service in Washington. Smith describes the major cooperative programs between the United States and other countries, and presents examples of successful ones. She also analyzes the failures or close failures, positing that a lack of program continuity, fluctuating policy toward technology transfer, and the question of assured launches play an important role in causing strife and disagreement among participants in a cooperative program. By attending to these problems and sources of conflict, we can develop programs that operate more smoothly. Nonetheless, as Smith rightly points out, the future of international space cooperation seems bleak; greater competition among the industrialized nations, politicization of resource exploitation issues, and growing East-West competition do not bode well for continuing cooperation.

Joseph Pelton looks at the problem of the proliferation of communications satellites in what is known as the Clarke orbit, that space 35,900 kilometers above the equator where satellites remain motionless relative to specific points on the ground. Pelton surveys the international system and emerging policy problems for the use of this orbit. The pivotal issue is guaranteed access, and he reviews the politics surrounding this problem. In contrast to Smith, Pelton is relatively sanguine about the prospects of technological and political resolution of the exploitation of the Clarke orbit.

Jürgen Häusler and Georg Simonis, both of Konstanz University in West Germany, examine how developed countries are using satellites that affect the Third World. The perspective of these two writers is highly pessimistic and in sharp contrast to Pelton's. They focus on the potential economic role of Germany's growing satellite industry. They see satellites as an extension of domination through technology and monopoly, and believe that satellites will harm, rather than help, developing countries. They also see that European countries will be banding together to help blunt the competitive technological advantage of the United States. While some may disagree with the value premises and argument that Häusler and Simonis present, they unquestionably raise provocative issues that are in need of further examination and that must not be ignored.

Jean-Louis Magdelénat discusses how satellites can be used to learn about conditions on, below, and above the Earth's surface. He then examines the political, economic, and especially legal ramifications of using satellites to collect and disseminate this information. After reviewing the major policy perspectives on these questions, he concludes that much greater effort should be devoted to international cooperation, and that national interests must increasingly be made subordinate to those of the international community. Magdelénat occupies a middle ground; he is deeply concerned about the potential conflicts involving remote sensing but, unlike Häusler and Simonis, does not look upon satellites as an exploitative tool. His position is probably closest to Pelton's, for both see the need for international organizations to evolve to manage the challenges of space technology.

Nathan C. Goldman of the University of Texas examines the difficult life of the Moon Treaty in his chapter. Now seemingly moribund, the intent of the treaty was to develop an international policy for the exploitation of lunar sources. However, the belief that the treaty would hinder resource recovery and harm the free enterprise system caused it great political difficulties both in the United States and in many other countries. With the change of administrations from President Carter to President Reagan, any limited, small hope that the United States would adopt the treaty was shattered. Goldman argues a treaty does not rule out the exploitation of lunar resources by private companies, but rather helps in the rational and organized tapping of such resources. He does not consider the treaty to be a threat to the free enterprise system.

In the book's third section, outer space is viewed as sociological phenomena. Six sociologists and a lawyer with an interest in sociology present their analyses. This section is an important step in bringing together views on the sociological implications of space.

The lead essay, by William S. Bainbridge of Harvard University, looks at the way individuals and social movements act to shape the mass movement of space exploration. Bainbridge's fundamental argument is that the venture into space has been driven by only a few clever individuals who were able to manipulate powerful political leaders. One of his conclusions is that space exploitation is, in a certain sense, a sham. He maintains that if it had not been for these creative men, there would be no "space movement" and no space policy today. Bainbridge further argues that unless new leaders come forth to manipulate political and societal forces, it is unlikely that there will be much progress in space exploration for the next several decades.

Bainbridge's analysis is followed by my chapter, in which I directly disagree with his position. My critique is directed at his theory, which is contained in his book *Space Flight Revolution*. I argue that internal societal forces lead to technological change and that important social forces, which can be neither stemmed nor induced by only a few persons, are responsible for the space program as it now exists. I review the literature on innovation to demonstrate my disagreement with Bainbridge.

Ron Westrum, David Swift, and David Stupple examine the obverse of what Bainbridge and I discuss. Rather than trying to understand the causes and implications of humans' reaching out to space and exploring the universe, these authors attempt to understand how human beings receive and perceive information about the possible existence of extraterrestrial beings. The authors survey the gamut of groups that believe that extraterrestrial intelligence exists, beginning with "contactee" groups who believe that they have had actual contact with visitors from outer space (and in some cases claim to have been kidnapped by aliens). They compare the norms and mores of such groups with those of scientists who are searching for extraterrestrial intelligence and those of groups believing in UFO phenomena. The points of agreement and conflict are delineated by the authors. They also examine the sociological processes through which these groups distinguish and distance themselves from one another.

B.J. Bluth of California State University–Northridge, intertwines studies of sociology, psychology, and human engineering to reflect on the problems and consequences of humans traveling in space. Her analysis is profoundly important in view of nascent plans to have permanent manned orbiting space stations. Her research should also be included in the body of literature taken into consideration before long-term manned planetary explorations are launched. She shows how the confined and isolated environment (as on a spaceship) contrasts with the ordinary environment in which people find themselves, a comparison that leads to several significant insights about human behavior in space.

George Robinson from the Smithsonian Institution looks at the legal ramifications of space "colonies," drawing parallels to the evolution of law and social control in the North American colonies and the westering frontier. Using this as a springboard, he speculates on how space colonies might evolve. Robinson is an advocate of manned exploration of space, and shows how law can both advance and hinder that exploration. He concludes, in contrast to Goldman, that space colonies will be imperialistic and exploitative rather than cooperative.

Several individuals have been immensely helpful to me in the editing of this book. I would like to note first of all the wonderful assistance I have received from Laura Battey. She played a pivotal role in helping me edit the book and was invaluable at every step of the production process. Irving Louis Horowitz inspired me to undertake this project and urged me to persevere. He early on identified space policy as an important area for social scientific analysis and has constantly urged that social scientists play a more vigorous role in analyzing, critiquing, and improving public policy. Mary Bauhs worked very hard on retyping chapters and has been a great help in the preparation of the manuscript.

This book has been many years in the making and the authors patiently endured several long delays and were generously willing to revise and update their chapters. I am grateful to each one. Their sharp and insightful analysis is a testimony to the contribution that the social sciences can make to understanding the policy process and the interaction between people and technology.

James Everett Katz

Introduction

1

American Space Policy at the Crossroads

James Everett Katz

In his 1984 State of the Union address, President Ronald Reagan committed the nation to orbit a permanent manned space station within the decade. Less than a year earlier, Reagan had proposed a multibillion dollar program to militarize space. The administration named the program the Strategic Defense Initiative (SDI), but it was quickly dubbed "star wars" by the media and public. The program's central goal is to create space weapons that can destroy either Soviet ICBMs attacking the United States or Soviet satellites orbiting the globe. Reagan's initiatives and the controversy swirling about them have reignited a keen interest in space programs on the part of U.S. policymakers, interest that had waned considerably, especially when one considers the level of interest that had existed during the period between Sputnik and the lunar landing (1957–1969).

This renewed attention to space casts in bold relief the policy problems that have cropped up during the period of space policy neglect. Interest, however, is concentrated on antisatellite and space antiballistic missile warfare rather than on the across-the-board use of outer space, even though technology has continued to advance in the latter area. The lag in policymaking presents problems, but also new opportunities, in the space policy field. (For a review of social science studies of space activity and policy, see the appendix to this chapter.)

With rare exceptions, like President Reagan's "star wars" speech, space policy over the past fifteen years has been made on an incremental basis by bureaus within the Office of Management and Budget (OMB), the Departments of State and Defense, and an isolated group of political appointees at NASA. Space enthusiasts in Congress are few in number, and their ardor has been matched only by their ineffectualness. Yet, the post-Apollo disenchantment with the space program should not dampen the interests of policy analysts; this relative neglect of the space program, combined with technological change, has created fertile ground for policy analysis that could influence the social scientific understanding of space policy and advance the processes involved in its formulation. Like the government, though, academic institutions also blow hot and cold on space policy. Not until the "star wars" speech did public policy analysts launch a flurry of activity.

Given the recent rise of space policy on the national agenda, it would be highly profitable to review the major space policy issues confronting the United States, and some of the options before U.S. decision makers. We will begin by examining NASA.

National Aeronautics and Space Administration (NASA)

NASA's rapid rise to success, culminating in the 1969 lunar landing, has been followed by a precipitous decline. In the late 1950s and until the mid-1960s, NASA was supported by the public and the media, as well as by Congress and the president. This policy "movement" exploited a powerful technology in the service of national prestige but, as some policy analysts such as Downs and Schulman[1] have suggested, when a program is quickly boosted to prominence by public support, it can just as rapidly fall into a state of neglect, disappear altogether, or become vulnerable to censure.

After the Apollo mission and with no new meaningful goal to take its place, NASA's lack of direction and tightening budgets began to neutralize organizational *esprit*, leading to problems of morale and the departure of superior personnel. The temporary end of manned missions had a widely depressive influence, and the diminishing enthusiasm of the space agency reduced organizational effectiveness.[2]

Along with an internal decline at NASA, the lack of tangible public support or funding has disrupted the associated network of university laboratories and training grants for space studies, and therefore the flow of newly trained space scientists. Aerospace contractors have also been hurt and many smaller firms have gone out of business, reducing the political support that can be marshaled from the private sector for a program in space. Sapolsky and Lambright have noted the efficacy of this "bleeding" technique for terminating research programs.[3]

In response to the worsening circumstances, NASA devised two strategies. The first was to maintain "steady-state" through the Apollo Applications Program, while a manned orbiting laboratory kept the technology alive at a minimal level until national interest could be rekindled.[4] Skylab met with public indifference, however, and the Apollo Applications Program, reduced enough to be acceptable to budget-conscious presidential staffs and congressional leaders, no longer sufficiently excited the public to mobilize support.

Failure may be attributable to a conundrum contained in the space policy situation: incremental policy necessitates modest programs, yet these fail to attract the political support necessary for approval.

NASA's other attempt to stay afloat involved the creation of imaginative and exciting goals that would rally public and political support for commitment to an ambitious space program. In 1969, a White House-level task force recommended a gigantic program including a moon base and a manned Mars mission.[5] Those plans were blocked because of their expense and because of public skepticism about the legitimacy of space exploration. Nevertheless, Nixon approved the Space Shuttle plan—on a greatly reduced scale—because lack of a significant project would spell the dissolution of the space program. Despite exciting launchings, the program is lagging far behind schedule, and because it is absorbing money that could otherwise be spent on planetary and space research, it has become a target for NASA's critics and for disaffected groups within and outside the agency.[6]

Some have blamed NASA's public relations department for its inability to deflect criticism and generate popular support, without recognizing that there are limits to the extent that public opinion can be manipulated by a single sector of society. Opinion polls consistently indicate that a large proportion of the American people believe a superabundance of capital has already been invested in space exploration, especially in the apparently wasted effort of reaching the moon. It is important to bear in mind that much of the early civilian space program was sustained by developments in military programs. The same knowledge needed for an ICBM force—about materials, electronics, guidance systems, computers, and reentry vehicles—was also necessary for the early space program, but today this overlap is much reduced. The space program itself was initiated and sustained by Cold War rationales. Considerations of national security and international competition placed the space issue beyond the control of any agency; NASA's decline is similarly the result of external factors beyond the manipulation of a single group. Still, attempts to stimulate public backing for space programs continue. One NASA suggestion was to ask the public to choose which of Jupiter's moons the spacecraft Galileo should visit on its planetary voyage, with the object of demonstrating that space programs are "public programs in the truest sense."[7]

President Carter was repeatedly asked to fill the space policy vacuum by setting and supporting NASA goals but only issued a lackluster policy paper in 1978 which made no commitments.[8] President Reagan has evinced a strong interest in space, but the indications are that only military space programs will be strongly supported. NASA budgets have been pared even more than under the Carter administration.

In a 1984 speech commemorating NASA's twenty-fifth anniversary, Reagan suggested that it was time for NASA to get out of the way and let private enterprise astound the world with what it could accomplish in space applications. This suggestion, in keeping with other as-yet unrealized suggestions by the administration for privatizing civil space activities, left NASA officials uncertain about their agency's long-term prospects.

With neither an influential constituency nor a presidential mandate, NASA is indeed vulnerable. Its attempts to promote exciting (and expensive) ideas usually attract criticism instead of a positive public response. Although ambitious projects involving power satellites or space colonies sometimes receive modest approval or publicity in Congress,[9] they are more likely to be subjected to scathing ridicule, as when Senator William Proxmire (D-WI) called the proposed space colony a "Beverly Hills in the Sky," and cautioned, "Not one penny for this nutty concept."[10]

NASA has been suffering an "identity crisis" due to the lack of a meaningful program and political support for the kind of goals that could revive interest. This situation shows the operation of reverse "technological obsolescence," or technology evolving beyond the political and the social system that manages it, instead of falling behind society's demands and uses for technology. Technical capabilities in space have outrun the ability of the nation to pay for them and of the political system to create and define purposes for the technology. An incrementally based decision–making system that prevents bold initiatives, and a changed focus for competition with the Soviets have left NASA with little mobilizing purpose. The 1970 characterization of the agency by then Senator Walter Mondale is still largely relevant: "What we see is a classic case of a program and agency in search of a mission."[11] Figure 1 indicates the degree to which U.S. space involvement has slowed relative to that of the Soviet Union.

Though the warming of relations with the USSR over space exploration reached a high point in 1975 with the Apollo-Soyuz docking, technological progress and doubt over Soviet global intentions has created a new climate of rivalry that is motivating another ambitious space program. This competition was evidenced when space program advocates succeeded in winning President Reagan's approval of a permanent manned space station. The $8 billion plan, announced in his 1984 State of the Union address, immediately divided Congress. The plan is ambitious indeed: by 1991, NASA expects to have an operational space station which can accommodate at least thirty people. If NASA and the president are able to persuade Congress to finance the program, NASA would have at long last the central focus necessary to sustain itself. But a space station is unlikely to capture the public's imagination in the same way that the lunar landing program did, and consequently will have to be strongly bolstered by a military rationale to claim a significant portion of the very tight federal budgets for the next decade. It is to this, the military side of space, that we next direct our attention.

FIGURE 1

Successful USA & USSR Announced Launches

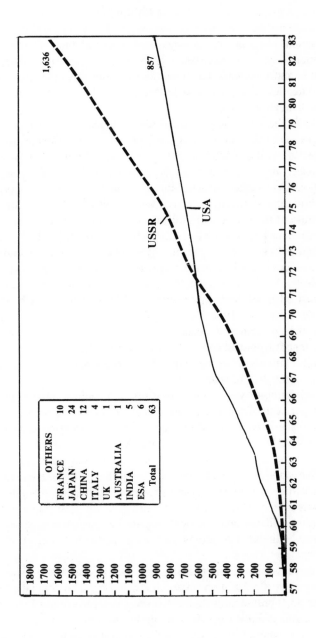

Source: NASA.

Military

National security has always been a key element in the nation's space program—now more so than ever. Satellites scrutinize military installations, radars, and telecommunications, and monitor compliance with arms limitations treaties, obviating on-site inspections. About 70 percent of U.S. military communications are transmitted by satellite.[12]

While the 1967 Outer Space Treaty banned weapons of mass destruction from space, military activities have continued there, though goals have shifted from "dropping atomic bombs" from satellites to combat between space vehicles, as in the current "star wars" scenario. (This chapter deals only with A-SAT warfare; implications of using space-based weapons against strategic nuclear deterrents, i.e. ICBMs, are discussed at length in chapter 4). Because of the reliance of both superpowers on satellites, the ability to neutralize them has become a crucial defense issue. If guidance satellites of the United States are destroyed, its seven hundred submarine-launched missiles become debilitated, and the threat to military satellites is no longer hypothetical. The Soviet Union has a satellite with intercept and attack ability against other satellites in low orbit and has conducted at least twenty hunter-killer tests since 1968.[13] Potentially, the Soviet Union's killer satellite system could be launched, rendezvous with, and destroy U.S. spacecraft in less time than it would take the U.S. satellite to make one orbit. While this system is still relatively crude and unreliable, it has sparked sharp controversy about a new arms race in space.[14]

U.S. satellite vulnerability is a result of policies and nonpolicies of the aerospace establishment. Because of satellites' cost effectiveness, many missions have been assigned to them without considering their long-term vulnerability. As a result, alternative or redundant systems have not been created. As expendable boosters are phased out, the United States will have only three to five shuttles for launching large objects, and these are easy targets for antisatellites (A-SATs). (The Air Force has recently begun a program to acquire new expendable boosters.) An optimistic assumption that the Soviet Union would not take the next step in the space war race, as well as ever-present budgetary limitations, have combined to hold back U.S. A-SAT technology.

To recover from this past neglect, the Department of Defense (DoD) recently spent 0.25 billion on A-SAT defenses. The official plan is to spend about $3.6 billion over the next several years to defend U.S. satellites from attack and to create a U.S. satellite-killing force. This may be a very conservative cost estimate; critics at the General Accounting Office contend that fulfilling the administration plan could cost $15 billion.

The primary A-SAT system being developed is a two-stage missile fired from an F-15 aircraft operating at high altitude.[15] It is capable of neutralizing a reconnaissance or other satellite in low orbit, but cannot intercept a Soviet A-SAT because the A-SAT's orbit cannot be calculated precisely enough. It also cannot reach satellites in high orbit. Lasers operating at the speed of light may solve these problems: the DoD has developed lasers that can destroy some planes and missiles.[16] The Pentagon is also researching the military applicability of particle beams.[17] The particle beam's utility as a weapon is largely unknown, but proponents claim the Soviets are building a particle-beam A-SAT weapon system to knock out incoming ICBMs. Significant efforts are also being devoted to the passive defense of satellites to reduce their vulnerability.

The most disturbing aspect of the increasing A-SAT activity is its political effect on the fragile but vital deterrent created by balancing world offensive and defensive power. Such balance would be destroyed by a sophisticated A-SAT system with its large military advantage, which could lead to a major escalation of the arms race and spread it further into outer space.[18]

New strategic doctrines are being developed that interpret the increase in A-SAT activities as the evolution of a truly automated battlefield. It is theorized that a satellite "cold war" and arms race would spare Earth from a nuclear Armageddon and involve only robots and remotely guided satellites.[19] However, disarmament specialist Herbert Scoville, Jr., maintains that an attack on a reconnaissance or communications satellite would be interpreted as a prelude to an all-out attack, thus leading the contestants irretrievably into terrestrial war. If conflict in space occurs, he says, "it will only be the first stage of a much broader war."[20]

Experts disagree about the motivations for the Soviet A-SAT programs. Some argue that it is a response to the U.S. Space Shuttle, which can intercept satellites and even return them to Earth for detailed inspection. Others maintain that the program generally reflects the reaction of a closed society to the constant scrutiny of outsiders, and specifically reflects the Soviet leaders' drive for secrecy. A third perspective argues that the program is not directed against the United States but, rather, at the more primitive Chinese reconnaissance system. The most widely accepted viewpoint is that the program is aimed at paralyzing the U.S. strategic command and control system in one bold stroke. The temptation of unilateral advantage offered by A-SAT technology has proven irresistible to the Soviet military establishment.

Whatever the program's true rationale, the Soviet pursuit of A-SAT technology has motivated the United States to push its own program. After several unsuccessful attempts by the United States to secure a treaty prohibiting A-SAT weapons, it now seems that the Soviet Union has become more receptive to such a treaty, due to increasing A-SAT research and development by the United States and the knowledge that U.S. technology could overtake the Soviet advantage in A-SATs within five years.[21]

In pursuit of world peace, numerous proposals have gone beyond the bilateral format typical of the A-SAT negotiations. However, complete neutralization of space, as some have recommended, seems unfeasible because the technologies often serve both military and peaceful uses and can readily be switched from one to the other. Moreover, many ''military'' uses of space strengthen world security and peace, and add to mutual deterrence capability. Still, the increase in A-SAT testing is an alarming development not prohibited by international law or agreement, and obviously leading to a frantic technology race. A treaty banning all weapons from space, and A-SAT testing in particular, seems justifiable.

Yet the U.S.-USSR bilateral approach excludes coordination with other countries that have their own perceptions and fears about the space arms race. As more nations attain launch capability, they too must be included in agreements designed to preserve space peace. Policies that protect ''freedom of passage'' for satellites in space have been motivated not only by military or security reasons but also by economic and political considerations. In this context, France proposed an international satellite-monitoring agency, and Italy advocated a United Nations-sponsored multilateral discussion on disarmament and outer space.[22]

Remote Sensing and Observation

Remote sensing technology began in the 1880s when cameras in balloons photographed the Earth by remote control. Today the Landsat series and military satellites belonging to the United States and other countries are constantly scanning the globe on both the visible and invisible segments of the electromagnetic spectrum. They can detect objects as small as 1 inch in diameter, temperature variations of .02 of a degree, trace amounts of minerals and chemicals, and a wide range of the electromagnetic spectrum (e.g., radioactivity, radar, and electronic communications).[23] This technology has profound implications for environmental quality, allocation and conservation of natural and social resources, and general scientific understanding; indeed, it has been compared to the microscope in its potential revolutionary impact. Remote-sensing technology offers great opportunities for decision makers to survey human and cultural resources, pollution emissions, plant disease, and building conditions. It also permits reliable estimation of troop strength and deployments, armament types, ship and aircraft dispositions, and weapons (including missile) testing.[24]

This powerful tool has understandably engendered international contention centering around two predominating issues. (See chapter 10.) First, United Nations General Assembly resolutions assert that peoples and nations have a right to permanent sovereignty over their natural wealth and resources, and that this right must be exercised in the interest of the national development and well-being of the people concerned. Several nations, including Argentina and Brazil, extend this right beyond the material wealth of the state to *information about* such wealth, and assert that unrestricted sensing activity of territorial resources violates sovereignty.[25]

The United States opposes this position, maintaining that the United Nations Charter, Article 19, advocates the right of all to "seek, receive and impart information and ideas through any media and regardless of frontiers." The United States has sought to reduce opposition to its position of comprehensive surveillance by giving any individual, entity, or state complete access to any data gathered by the Landsat series upon payment of a nominal fee. This access is extended even to nations with hostile political philosophies, such as the Soviet Union and the People's Republic of China.[26]

The U.S. position is weakened, however, by numerous court decisions and treaties here and in other countries giving a state the right to control or prohibit photography in its airspace or even photography *into* its territory, though no definition of "airspace" is universally accepted. Some states would extend the term "airspace" to include the Landsat orbits, just as territorial rights have been said to extend into coastal waters. Both claims have traditionally been limited by the available technologies to enforce them.

For obvious reasons, remote data gathering presents economic, legal, political, and security threats. Satellites can easily provide economically useful data, such as crop conditions, location of ore deposits, or factory construction. Foreknowledge of agricultural conditions and thus of commodity prices can affect the economic viability in the world market of some states that are heavily dependent on sales of certain agricultural products. Remote sensing technology can also give an unfair advantage in negotiating contracts for exploration or exploitation rights to those who hire experts to interpret satellite-generated data. Although U.S. courts have consistently maintained that photographs and information are protected property rights, this nonphysical, but still exchangeable form of property is daily expropriated by the government through satellite sensing. Although no one has litigated this issue, legal proceedings have been initiated on the basis of evidence gathered by satellites, as when water-borne pollution was detected by Landsat and a court battle over violation of pollution laws resulted. This is the first of what could become a host of legal cases related to satellite-generated data, involving zoning laws, personal privacy, criminal conduct, and industrial espionage.[27] The legal and constitutional issues of privacy, evidence, and data gathering that result from remote sensing have only begun to be studied.[28]

Many nations tightly control even innocuous data on resources, population distribution, and level of activities, perceiving its release as a needless threat to their stability and security. Other countries' objection to data gathering about military activities is more understandable, considering that satellites are able to read license plates and even insignia on uniforms.

Resistance to satellite observation has resulted in a series of countermeasures which attempt to baffle satellite reconnaissance. Many such ruses, like the Soviet "cold belts" buried in the soil to disguise infrared radiation from underground air bases,[29] have failed, but it is impossible to tell how many have succeeded. Nevertheless, remote sensing has become the keystone upon which strategic arms limitation is built because it eliminates the necessity of on-site inspections yet permits the monitoring of treaty compliance.

The Soviets, later joined by the French, have advanced a plan whereby remote-sensing data cannot be made public or "communicated to third parties . . . without the consent of the state whose territory is affected." Argentina, later joined by other Latin American countries, sponsored an even stronger position, prohibiting any remote sensing activity without prior consent of the surveyed nation. The Soviet-French proposal merely prevents transfer of the data to third parties; the Latin Americans would completely prohibit data gathering.[30] Meanwhile, the United States has consistently supported a position of open sensing of the world's natural resources and unrestricted release of all data derived therefrom. It seems that a de facto if not de juris compromise is being reached whereby the governments that gather the data will freely release them, but will refrain from publishing interpretations about the natural resources of other nations. Nonetheless, continuing advances in remote sensing technology will increasingly invoke questions of privacy, property, political control, and national security. U.S. policymakers will be aiming at protecting personal and property rights, without blunting entrepreneurial initiative; controlling global resources more efficiently, while avoiding unfair exploitation of the Third World; and recognizing the apprehensions of less secure nations, while encouraging their openness to sensing.[31]

The Geostationary Orbit

The geostationary orbit, or Clarke orbit, is a corridor at about 35,900 kilometers above the Earth's equator where satellites can be placed and maintain a constant position relative to Earth. This orbit is an ideal parking place for a satellite in constant contact with a particular ground station. However, the number of segments of the orbit which can be occupied by any particular satellite is limited; the orbit is already congested over the United States and

Canada. With increasing demand for telecommunications, broadcasting, meteorological, and possibly power transmissions, controversy has intensified over use of the geostationary orbit.

The United States maintains that space can be freely explored and used by all states without discrimination, if the purpose is for the "benefit and in the interests" of all countries. Its disinclination to limit the definition of "benefits," however, has led to increasing dissension. After claiming a segment of the geostationary orbit above its national territory, Colombia, and seven other equatorial countries (Brazil, the Congo, Ecuador, Indonesia, Kenya, Uganda, and Zaire) issued the Bogota Declaration, which asserts that each country's segment is a natural resource and an integral part of its sovereign territory. As foreign occupation of these segments would be tantamount to appropriation of sovereignty and resources, the Bogota Declaration seemed to countervene directly the 1967 Outer Space Treaty. Holding that the treaty does not apply to the geostationary orbit, the equatorial countries defended their declaration on the basis of justice, equity, and fairness rather than international law. They argued that the current practice, supervised by the International Telecommunication Union (ITU), will leave the developing nations with no segments of the geostationary orbit or with the least desirable locations and frequencies. Thus, the countries whose financial and technological resources are the most modest will have to spend the most. The Bogota Declaration is also defended by the argument that the 1967 treaty, as applied, does not carry out the intention of its drafters, which was that space development benefit all nations to achieve a more just international order.

The Bogota group's scientific and legal arguments are given little heed by the rest of the world. The United States views the declaration as appropriation of outer space. Some nations have suggested privately that additional sources of revenue, rather than concern about the international economic order, motivated the declaration. Meanwhile, U.S. use of the Clarke orbit proceeds apace. A U.S. commercial satellite operator recently placed one of its satellites in geostationary orbit several years before its service will be needed, in order to preempt the space. Western Hemisphere nations met in 1979 to seek equitable distribution of the five available satellite spaces only to discover that the United States already occupied four of them.[34]

One proposal for such problems suggests splitting the geostationary corridor into sectors and allocating one to each nation. Understandably, the United States would resist a reduction in its present control over many segments and, in any event, the proposal would be likely to cause as many problems as it would solve. No specific plans effectively address the current situation, which is deadlocked because the equatorial nations cannot remove or interfere with the stationed satellites. If they do attempt to enforce their claims, terrestrial action against alternative targets, such as an embargo or property seizure, is possible. The United States should develop policy options to deal with such a contingency.

The exploitation of space has not followed any predetermined and rationally developed plan but has evolved gradually. Following the Soviet example with Sputnik, no nation has asked permission before orbiting a satellite, and no official protests were heard until Colombia's in 1975. The first-come, first-served mode of operations in space took its rationale from the principle of free exploration and use of outer space, and became ratified in the 1967 Outer Space Treaty. Latecomers have no recourse but to make claims against those who are already using a particular space resource. The United States has always placed efficient, economic use of space before principles of equitable access and general sharing of benefits, and its superb technology has permitted full exploitation of this approach. However, in recent years, the less developed nations have been asserting their rights with policies that would allow them a greater portion of the world's material benefits.[35] Their new assertiveness is manifested in negotiations on seabed and lunar exploitation treaties and in the results of United Nations resolutions and meetings.

The challenge is to devise a policy for the orderly allocation and use of geostationary orbits, one which will also encourage the development of space resources, so that the interests of all countries will be recognized and protected.

Direct Broadcasting

The international controversy over communciations satellites includes communciations themselves as well as positioning. Communications satellites can broadcast into individual home receivers, directly challenging some governments' control.[36] For political, cultural, and economic reasons, some countries want to monitor information or programs broadcasted into citizens' homes, including "cultural" programs that may be distasteful or contain hostile overtones, unwelcome information, or invite invidious comparisons. Even "open" societies may resist direct broadcasting by satellite in fear that their national culture could be submerged. Many nations find that, although U.S. commercial programming is visually attractive, it is generally "trivial, banal and violent."[37]

The United States along with some other nations has maintained that the flow of data and information should be subject only to minimal regulation. Conceding that prior notification and consultation should be required, it continues to resist guidelines that it believes inhibit future technological development and the free exchange of ideas and information (and markets for U.S. broadcasters). (For a critical review of the impact of satellites on Third World countries, see chapter 9.)

Predictably, the USSR advocates prior consent, controlled program content, and the prevention of illegal broadcasts. Argentina's position is similar but with added restrictions: a right of reply, and proscriptions against any tele-communication that threatens the security of the state or the family, or impairs the rights of families or individuals, or is contrary to public law or morality.[38]

These proposals leave three unresolved policy issues in the area of direct broadcasting from satellites: the free flow of information; recognition of the value of cultural diversity; and a nation's right to determine the character of the television services available to its people. The resolution of one is likely to increase conflict between the others.

Exploitation of Space Resources

International policies for the exploitation of space resources have generated the same concerns as the Law of the Sea negotiations concerning mining of the seabed. Essentially, the nations without technological capability want to share nonterrestrial resources that they believe are the common heritage of all humankind. The technologically advanced nations argue that this sharing of benefits would destroy the commercial incentive to exploit space resources. As plans to mine the moon or asteroids have circulated, some nations have pressed for an international policy to govern the exploitation of celestial resources.

In 1970, Argentina presented the first draft treaty on the use of the moon to the U.N. Committee on Peaceful Uses of Outer Space (COPUOS). Initially it was largely ignored because the Outer Space Treaty of 1967 was thought to deal adequately with the moon and other celestial bodies. But over time, it gathered advocates and was seconded by the United States, India, and Egypt.[39] The draft treaty was passed by COPUOS and the U.N. General Assembly and, by 1984, was ratified by four countries and signed by seven others. Although the Carter administration signed the treaty, the Senate refused to ratify it, and the Reagan administration withdrew it from further consideration. The treaty, emphasizing peaceful uses of space for the benefit of all, mandates that any resource development include the interests of present and future generations and the promotion of higher living standards and social and economic progress, especially in less developed countries. It states that the moon and other celestial bodies must be used exclusively for peaceful purposes and prohibits weapons of mass destruction, testing of weapons, military installations, and maneuvers. It assures free access to the moon, which includes the full use of any lunar facility or equipment to astronauts whose lives are endangered. (See the Goldman chapter on the moon treaty in this volume.)

The fundamental disagreement over the treaty has involved natural resources and the international arrangement necessary to regulate their exploitation. The original United States position, embodied in the treaty, favors (1) orderly and safe development of the moon's natural resources; (2) rational management of the resources; (3) expansion of opportunities in the use and exploitation of the resources; and (4) equitable sharing of benefits derived from the resources. The treaty does not define "equitable sharing of benefits." Some say the treaty implicitly places a moratorium on resource exploitation in space until an international authority is established to control it, but the United States denies this. Others fear that the international authority model devised to control (and which effectively prevents) seabed mining will be applied to the exploitation of celestial resources. It has been opposed by several domestic interests, such as industries contemplating the exploitation of lunar resources who see the treaty as an impediment. The American Mining Congress has, in fact, testified to the Senate that the treaty was contrary to U.S. interests and that it may have the same effect as the Law of the Sea, which was to extinguish corporate interest in seabed mining through international sharing and control provisions. If this is so, the treaty would delay the creation of needed technology and the exploitation of both the physical resources of space and of solar power from satellites.

Treaty advocates hold that the international regime need not blunt lunar exploitation as the sea authority has marine mining. They point to the International Telecommunications Satellite Organization (INTELSAT) as an example of companies jointly owning and operating a lucrative international system. Furthermore, the moon agreement states that a regime is necessary only when exploitation of space resources becomes feasible and practicable, which may be years away, or never. In the meantime, institutional arrangements can be created for rational experimentation and gradual development of effective procedures. Finally, advocates say the United States has left itself an escape mechanism by committing itself only to work on the establishment of an international regime, not necessarily to implement one or participate in one. But some legal authorities and those critical of the treaty assert that the treaty makes an international regime a *sine qua non* of resource exploitation.[40]

On the level of practical politics, the United States may have originally accepted the restrictive language of the treaty to win concessions from other nations on issues considered more vital to its interests than those addressed by the treaty. Remote sensing and geostationary orbits (called "freedom of space passage and navigation") and direct broadcasting ("freedom of information") are considered more important than the distant prospect of lunar mining. Moreover, the United States may have acted to fill a vacuum that could otherwise attract concepts of an international regime even more hostile and contrary to its interests than the one now being implemented. Still, below this level of rationalization is the fear that U.S. policy on space resource exploitation resulted from incompetence, i.e. inattention by top policy officials, and a lack of recognition of the magnitude of the issues involved. Some say any U.S. negotiator familiar with the international seabed mining agreements would have avoided supporting a moon treaty like the present one. It has been privately expressed that the handling of the treaty issue was marked by mistakes and failures from which the United States cannot escape without significant damage to its credibility and prestige.

But beyond political and economic vagaries, the crux of the issue is the development of an institution to create and implement efficacious space policies that are agreeable to all parties. Clearly, opinions are divided about the soundness of the approach used in the moon treaty and especially the ideas it embodies about an international regime. It is generally agreed, however, that the opening of space presents new organizational opportunities and challenges in the search for effective and just international systems. Options for the institutional management of space policy are considered in the next section of this chapter.

Institutional Frameworks for Space Law

The growth in possible uses of outer space has resulted in a search for institutional arrangements that can best guide space exploitation. Pivotal issues in the search are the criteria and procedures to determine access to space, how outer space resources should be exploited, and to whose benefit. The previous discussions of numerous conflicts suggest not only that the inherited international regime is inadequate for outer space but that it is becoming increasingly problematical in general. Because technology has developed more rapidly than social institutions, nations do not agree about the best means of efficiently and fairly allocating resources. Theories generally originate in one of two approaches to the problem. The first, the "national authority" approach, is the result of nation-states assuming responsibility and authority for the allocation of resources and, where possible, sovereign control. The second, "functional eclecticism," is the incremental growth of limited international authority, resulting in an ungainly mixture of specialized functions allocated to several institutions, and many general issues unresolved.[41]

The national authority approach, the predominating response to the exploitation of outer space, essentially extends the traditional land-oriented philosophy of resources and also relies on the fact that national governments are convenient institutions that claim to act in the public interest, in either national or international terms. The uses of space are and have been exploited primarily by nations. The International Telecommunication Union (ITU) and the International Telecommunications Satellite Organization (INTELSAT) have been the main exceptions. As indicated earlier, the space-faring nations argue that their exploitation of the common resource of space is based on the rationale of efficiency. They retain control over the creation, deployment, and management of not only their national program but international space service programs (such as Landsat) as well. These states, primarily the United States and USSR, argue that because space applications require such vast expenditures for research and development, operations, and facilities maintenance, neither the public nor private sectors would be willing to underwrite the projects if the benefits were more widely distributed.[42]

It has been countered that this nationalistic response is basically retrogressive, since states, by their nature, are insensitive to the social and economic interdependence that characterizes the contemporary world. The use of national systems for international problems aggravates the contradictions between the real qualities of an interdependent world and the traditional international political system. Also, it is argued that permitting nations to exercise authority beyond their scope and competence will undermine their legitimacy.[43]

The issue of direct broadcast, discussed earlier, depicts how national authority regimes are unable to safeguard the concerns of other nations. In response, nations viewing themselves as victims of "technological imperialism" by direct broadcasting could, justly or not, mobilize international coalitions to restrict activities in space which otherwise would be of substantial benefit to a particular nation, the world in general, or even to a developing country.

The usual approach to solving such problems is to create international organizations with responsibility over particular ones, while the general area is left to the nation-states to be handled as they see fit. While conflict avoidance and enhancement of international reputation have predominated as motives for the development of international agreements and agencies, in the future economies of scale may also be a major factor. This will most likely proceed as in the past, i.e. incrementally, experimentally, and topically, which tends to maximize the flexibility of organizational arrangements. Nevertheless, it is unlikely that leading space nations will cede power over crucial areas like security affairs to a decision-making body responsive to demands of a larger international constituency. Indeed, it is surprising that the United States may be willing to give an international body control over the potentially great wealth of lunar resources; such a concession, as was pointed out, would be made with domestic resistance.

One of the greatest benefits of the incremental approach to international law frameworks is that it eliminates the necessity of trying to bring about an all-or-nothing grand design for space law, which could paralyze the international decision apparatus and cause sharp ideological disagreements between nations and coalitions. On the other hand, the complex interaction of interests, coalitions, institutions and agreements not only is confusing but diffuses political accountability for the use and abuse of outer space. This can be seen, for example, in the controversy over the allocation of geostationary satellite orbits. Crisscrossing of authority and programs enables corporate and state entities to avoid meaningful accountability to a higher national or international public interest.[44]

As an experimental and gradual approach, "functional eclecticism" is appropriate to situations in which the directions of social and technical development are unknown. It does not force structural unity on an inherently diverse and inchoate international social structure. However, it also circumvents accountability to a broader public interest and need, is narrowly unilateral, motivated solely by self-interest, and lacks the vision necessary for a world community.[45]

An obvious institutional solution would be to merge the flexibility of eclectic functionalism with a transitional strategy to establish a powerful agency for outer space projects. Such an agency, with an international membership, could gradually internationalize capabilities for gathering and assessing most information derived from remote sensing and information about outer space, leading to a fuller sharing of these data. It would also push for intensive international consultations to resolve contending interests and to produce a wider range of policy choices for the international community. Finally, such an agency might limit nationally oriented activity in outer space by replacing it with binding reciprocal obligations that would lead to a mutual accountability network.

The attempt to establish an outer space agency would be challenged by national interests that might have to forfeit power and control to it, or that see it as a threat to national well-being. It also may be resisted by less developed countries that might see the agency as perpetuating an inequitable situation. But the moon treaty suggests that the agency could come into being, especially if it could handle policy issues that have not yet become of public interest: the mining of resources on Jupiter, for instance. Starting in new areas would counter the paralyzing effect of prior institutional barriers and would allow enough flexibility to incorporate technological advances effectively.

Conclusion

This overview of emerging space policy problems has intentionally shifted across many different levels of analysis to show the abundance of issues of concern to policy analysts. The field reveals a rich quarry for those interested in bureaucratic politics, decision-making models, organizational processes in constrained environments, links between public opinion and public policy, links between international and domestic policy, societal development, and even futurology. Moreover, the general neglect of the space policy area has reduced the cacophony of "commonsense" rhetoric from politicians and ideologues, leaving a zone relatively free of politics and entrenched interests into which social scientists can venture with some chance of affecting policy. But this zone will not long remain tranquil. The "winds of star wars" have begun to blow. As their velocity increases, social scientists will have less and less opportunity to make a difference in policy formulations.

Appendix

In contrast to the relative oblivion to which it is has been relegated in recent times, national space policy received concentrated attention from the policy analysis community during its foundation. Numerous studies examined the rationale behind the early space program and the decision to go to the moon.[46] Advocates of the "bureaucratic politics" decision model found the military's rocket programs a rich area to explore; various "crisis" models have been applied to the Apollo decision.[47] The space program has been examined in terms of national political processes[48] and distributional politics,[49] and space research has been framed in a grand scheme of science policy analysis.[50] NASA has also served as a focus of internal organizational studies, with the purpose in some cases of developing recommendations for structuring high-technology, rapidly changing organizations based upon the NASA experience.[51] Economists, most sponsored by NASA or its contractors, have scrutinized NASA's spending to show that there is a significant return on investments,[52] and their studies have in turn been criticized.[53] NASA programs have been examined from various perspectives to demonstrate their impacts on local communities and industrial groupings, especially the aerospace sector,[54] and NASA's technology transfer programs were evaluated.[55] The tone of most of the program analyses has been positive, but some scholars have vigorously attacked NASA management and programs.[56]

Investigations of the broader social impacts of NASA's activities include the modest attempt in the macro- and micro-sociology of science undertaken by James Kuhn and Ian Mitroff, respectively.[57] Bauer and others investigated the second-order consequences of the space program and sought to use the technological transformation of the American railroad industry as a methodological analogue for changes that might accompany a large-scale space program.[58] The result of this research merely demonstrates the inadequacies of argument by historical analogy.

Few serious social scientific analyses of space policy have been published in recent years, and most of these have been written by lawyers in international and space law[59] or by military officers interested in strategy and national security.[60] They leave many interesting issues in need of careful policy analysis from a social scientific perspective. The most recent published policy study of NASA, for example, appeared in 1975.[61]

Society's interest in space and the evolution of the social underpinnings of astronautics have been studied from the perspective of social movements and by extending Thomas Kuhn's theories of social change.[62] While provocative, this approach has serious shortcomings and has thus attracted criticism.[63] The social and economic bases of life in outer space, either in floating space colonies or on the surface of celestial bodies, have in recent years been the subject of constant speculation in traditional scientific journals and in futurics-style publications.[64] Space manufacturing facilities and especially solar power satellites have been evaluated in economic, engineering, and social terms,[65] and, while there seem to be no technological barriers to their realization, it appears highly unlikely that the exorbitant amounts of money needed for space construction will be available soon.[66]

The psychological conditions and consequences of living in space have been examined by specialists, and research has included simulations.[67] (See chapter by Bluth in this volume.) The sociopsychological reaction to space exploration has also been investigated (see chapter 14).

Because most studies of the social and economic bases of space habitats extrapolate from contemporary values and economic facts, their future validity is questionable. The subject begs for careful, empirically based, multidisciplinary inquiry. NASA officials contend that they are willing to support such research but have not yet been presented with scientifically defensible proposals.

Notes

1. A. Downs, *Inside Bureaucracy* (Boston: Little, Brown, 1967); P. Schulman, "Nonincremental Policy Making," *American Political Science Review* 69 (1975): 1354–70.
2. M. Mueller, "Trouble at NASA," *Science*, 22 August 1969, pp. 776–79.
3. W. Lambright and H. Sapolsky, "Terminating Federal Research and Development Programs," *Policy Sciences* 7 (1976): 199–213.
4. For a description of these programs, see E. Redford and O. White, "What Manned Space Program after Reaching the Moon?" (Syracuse: Inter-University Case Program, 1971); U.S. Senate, Science and Technology Committee, *World-Wide Space Activities* (Washington, D.C.: Government Printing Office, September 1977), pp. 31–87.
5. U.S. Space Task Group, *The Post-Apollo Space Programs: Directions for the Future*, Report to the President (Washington, D.C.: Government Printing Office, 1969).
6. R. Smith, "Shuttle Problems Compromise Space Program," *Science*, 21 November 1979, pp. 910–14; A. Large, "America's Space Shuttle Lemon," *Wall Street Journal*, 31 January 1980, p. 14; U.S. House of Representatives, Science

and Technology Committee, *Hearings: Space Shuttle Operational Planning Policy and Legal Issues* (Washington, D.C.: Government Printing Office, 1980).

7. R. Smith, "Uncertainties Mark Space Program of the 1980's," *Science*, 14 December 1979, p. 1286. A public opinion survey on Project Apollo found most Americans *did not* think the lunar landing was worth the effort; *New York Times*, 20 July 1979, p. 12.

8. "United States Space Activities," *Weekly Compilation of Presidential Documents* 14 (26 June 1978): 1135–37.

9. U.S. General Accounting Office, "U.S. Must Spend More to Maintain Lead in Space Technology," FGMSD-80-32 (Washington, D.C.: 1980); U.S. Senate, Commerce Committee, *Hearings: U.S. Civilian Space Policy* (Washington, D.C.: Government Printing Office, 1979).

10. K. Maize, "Changing U.S. Space Policy," *Editorial Research Reports* 2 (10 November 1978): 847.

11. J. Katz, "Chariots and Drivers: Social Forces and Technological Change in Space Policymaking," *Futurics* 5 (Spring 1981): 97–105.

12. R. Toth, "War in Space," *Science 80* (September 1980): 76; Thomas Krebs, "The Soviet Space Threat," *Journal of Social, Political, and Economic Studies* 9 (Summer 1984): 144–63; Jean-Pierre Clerc, "The Militarization of Outer Space," *World Press Review* 30 (October 1983): 23–25.

13. *Wall Street Journal*, 23 February 1984, p. 22; *New York Times*, 21 May 1981, p. A26. The A-SATs are effective only against near-space objects. The high-altitude geosynchronous satellites, such as NAVSTAR, appear to be beyond the capability of Soviet A-SATs for at least another decade.

14. See chapter 3 in this volume for more on this point. See also Keith B. Payne and Colin S. Gray, "Nuclear Policy and the Defensive Tradition," *Foreign Affairs* 62 (Spring 1984): 820–42; William Burrows, "Ballistic Missile Defense," *Foreign Affairs* 62 (Spring 1984): 843–56.

15. *New York Times*, 5 June 1983, p. D-5. See also *Aviation Week*, 25 June 1979, p. 25; 3 September 1979, p. 57.

16. B. Thompson, "Directed-Energy Weapons and the Strategic Balance" *Orbis* 23 (Fall 1979): 697–709; *Aviation Week*, 8 October 1979, p. 15; 28 July 1980, pp. 33–66; 4 August 1980, pp. 44–68.

17. *New York Times*, 5 June 1983, p. D-5. See also the testimony of Major General Donald Lamberson, "DoD Directed Energy Programs," U.S. House of Representatives, Armed Services Committee, *Hearings: Department of Defense Authority and Oversight for FY 84, Part 5* (19 April 1983); T. Bell, "America's Other Space Program," *The Sciences* (December 1979): 12; J. Mason, "Particle Beam Weapons: A Controversy," *IEEE Spectrum* 16 (June 1979): 30–35. For a critical view of particle beam weapons, see J. Parmentola and K. Tsipis, "Particle Beam Weapons," *Scientific American* 240 (April 1979): 54–65; P. Laurie, "Exploring the Beam Weapon Myth," *New Scientist* 82 (26 April 1979): 248–50.

18. *U.S. Policy on A-SAT Arms Control* (Washington, D.C.: The White House, 31 March 1984); Michael R. Gordon, "Proposed U.S. Anti-Satellite System Threatens Arms Control in Space: Reagan's 'Star Wars' Proposals Prompt Debate Over Future Nuclear Strategy," *National Journal*, 7 January 1984, pp. 12–17; Ashton Carter and David Schwartz, eds., *Ballistic Missile Defense* (Washington, D.C.: Brookings Institution, 1984). See also chapter 4 of this volume.

19. Stockholm International Peace Research Institute, *Outer Space: Battlefield of the Future?* (London: Taylor & Francis, 1978), pp. 167–88; *Aviation Week*, 3 September 1979, p. 57.

20. Toth, "War in Space," p. 80.

21. *Science*, 18 May 1984, p. 695; *Washington Post*, 28 January 1984, p. A–23; *New York Times*, 19 August 1983, p. A–3; 30 August 1983, p. A-21. See also *New York Times*, 11 June 1978, sec. 4, p. 3; 10 April 1979, p. 12.

22. F. Lay, "Nuclear Technology in Outer Space," *Bulletin of the Atomic Scientists* (September 1979): 27–31.

23. Bell, "America's Other Space Program," p. 8; S. Barber, "Watching Brief on the World," *Far Eastern Economic Review* 95 (25 February 1977): 26–29.

24. H. de Saussure, "Remote Sensing by Satellite," *American Journal of International Law* 71 (October 1977): 707–24; D. Jordan, "Looking in on Us," *Environment* 19 (August 1977): 6–11.

25. J. Hahn, "Developments Toward a Regime for Control of Remote Sensing from Outer Space," *Journal of International Law and Economics* 12 (1978): 421–58; N. Hosenball and P. Hartman, "Dilemmas of Outer Space Law," *American Bar Association Journal* 60 (March 1974): 298–303.

26. U.S. House of Representatives, Science and Technology Committee, *Hearings: International Space Activities* (Washington, D.C.: Government Printing Office, 1978).

27. De Saussure, "Remote Sensing by Satellite."

28. Jordan, "Looking in on Us"; D. Campbell, "Threat of Electronic Spies," *New Statesman*, 2 February 1979, pp. 142–45. The *New York Times* reports remote sensing satellites were used to monitor some Vietnam War protests in the U.S.; 17 July 1979, p. 10.

29. Barber, "Watching Brief on the World," p. 27.

30. De Saussure, "Remote Sensing by Satellite."

31. See A. Chayes et al., "A Surveillance Satellite for All," *Bulletin of the Atomic Scientists* (January 1977): 7.

32. Hosenball and Hartman, "Dilemmas of Outer Space Law"; C. Busak, "The Geostationary Orbit: International Cooperation or National Sovereignty?" *Telecommunication Journal* 167 (1978): 49–74.

33. S. Gorove, "The Geostationary Orbit: Issues of Law and Policy," *American Journal of International Law* 73 (July 1979): 441–51. See also, J. Pelton and M. Snow, eds., *Economic and Policy Problems in Satellite Communications* (New York: Praeger, 1977).

34. D. Moralee, "INTELSAT: Facing the Problems of Success," *Electronics and Power* (November 1979): 778; *New York Times*, 15 January 1979, sec. 4, p. 2.

35. For a perspective in this process, see I. L. Horowitz, "Death and Transfiguration in the Third World," *Worldview* 20 (September 1977): 20–25; K. Wolf, "Conflicts and Cooperation in the Opening-Up of New Economic Resources: The Third U.N. Conference on the Law of the Sea," *Intereconomics* 15 (January 1980): 21–28.

36. P. Dauses, "Direct Television Broadcasting by Satellite and Freedom of Information," *Journal of Space Law* 3 (1975): 59–63; Pelton and Snow, *Economic and Policy Problems in Satellite Communications*.

37. A. Chayes and P. Laskin, *Direct Broadcasting from Satellites* (Washington, D.C.: West, 1975); United Nations, "Draft International Convention on Direct Broadcasting by Satellite," Document A/AC.105/134 (5 July 1974).

38. Mahdi Elmandjra, "The Conquest of Space," *Third World Quarterly* 6 (July 1984): 576–603; C. Horner, "Outer Space and Earthly Politics," *American Spectator* 12 (February 1979): 11–14.

39. G. Reijnen, "History of the Draft Treaty on the Moon," in *Proceedings of the 19th Colloquium on the Law of Outer Space*, ed. M. Schwarts, pp. 358–65 (Hackensack, N.J.: Rothman, 1977), p. 359; Y. Kosolov, "Legal and Political Aspects of Space Exploration," *International Affairs* (Moscow; March 1979): 86–92; W. Broad, "Earthlings at Odds over Moon Treaty," *Science*, 23 November 1979, pp. 915–16; K. Teltsch, "Pact on Moon's Riches Approved," *New York Times*, 4 July 1979, p. 4.

40. Broad, "Earthlings at Odds Over Moon Treaty," p. 916. Space colony supporters see their plans for developing space factories, power satellites, and habitats also endangered. The 3,500-member L-5 Society hired a lobbyist to stop passage of the moon treaty because it threatened to postpone indefinitely their plans to establish space habitats, and because they believe it constituted a fundamental abrogation of civil rights. One part of the treaty allows any government to inspect a space station in orbit around any celestial body except the Earth. A government would not need permission to board and search any space station as thoroughly as it wished. In sum, opponents of the treaty see it as an international instrument that could prevent space settlement and exploration, and slow, if not halt, the development of space technology and the exploitation of vital resources that would elevate world living standards. S. Brown and L. Fabian, "Toward Mutual Accountability in the Non-terrestrial Realms," *International Organization* 29 (Summer 1975): 877–92.

41. S. Socol, "Comsat's First Decade," *Georgia Journal of International and Comparative Law* 7 (1977): 678–92; S. Levy, "INTELSAT: Technology, Politics and the Transformation of a Regime," *International Organization* 29 (Summer 1975): 655–80. See also A. Krause, "Europe's Space Shot," *Europe* (January 1980): 12–13.

42. M. Olson, *Logic of Collective Action* (Cambridge: Harvard University Press, 1971).

43. Brown and Fabian, "Toward Mutual Accountability in the Non-terrestrial Realms," p. 880.

44. Hosenball and Hartman, "Dilemmas of Outer Space Law."

45. See United Nations, Department for Disarmament Affairs, *The Implications of Establishing an International Satellite Monitoring Agency* (New York: United Nations, 1983).

46. V. Van Dyke, *Pride and Power: The Rationale of the Space Program* (Urbana: University of Illinois Press, 1964); J. Goldsen, ed., *Outer Space in World Politics* (New York: Praeger, 1963); E. Schoettle, "The Establishment of NASA," *Knowledge and Power*, ed. S. Lakoff, pp. 271–92 (New York: Free Press, 1966); L. Swenson et al., *This New Ocean: A History of Project Mercury* (Washington, D.C.: NASA, 1966); M. Snow, *International Commercial Satellite Communications* (New York: Praeger, 1976).

47. T. Greenwood, *Making the Mirv* (Cambridge, Mass.: Ballinger, 1975); E. Beard, *Developing the ICBM: A Study of Bureaucratic Politics* (New York: Columbia University Press, 1976); E. Bottome, *Missile Gap: A Study of the Formulation*

of Military and Political Policy (Cranbury, N.J.: Fairleigh Dickinson, 1971); J. Logsdon, *The Decision to Go to the Moon* (Cambridge: MIT Press, 1976).

48. J. Katz, *Presidential Politics and Science Policy* (New York: Praeger, 1978); P. Schulman, "Nonincremental Policy-making," *American Political Science Review* 69 (December 1975): 1354–70.

49. T. Murphy, *Science, Geopolitics and Federal Spending* (Lexington, Mass.: Heath-Lexington, 1971).

50. D. Schooler, *Science, Scientists and Public Policy* (New York: Free Press, 1971).

51. P. Schulman, "The Reflexive Organization," *Journal of Politics* 38 (November 1976): 1014–23; H. Anna, *Task Groups and Linkages in Complex Organizations: A Case Study of NASA* (Beverly Hills: Sage, 1976): G. Berry, "NASA's Work Planning and Progress Review," *Civil Service Journal* (April 1979): 22–25; L. Sayles and M. Chandler, *Managing Large Systems* (New York: Harper & Row, 1971).

52. Chase Econometric Associates, Inc., "The Economic Impact of NASA R&D Spending," NASA-2741 (April 1975); Rockwell International Space Division, "Impact of the Space Shuttle Program on the California Economy," FD-74-SH-0334 (December 1974); Mathematica, Inc., "Quantifying the Benefits to the National Economy from Secondary Application of NASA Technology," NASA-2734 (June 1975).

53. Z. Griliches, "Issues in Assessing the Contribution of Research and Development to Productivity," *Bell Journal of Economics* 10 (Spring 1979): 92–116.

54. R. Konkel, "Space Employment and Economic Growth in Houston and New Orleans, 1961-1966" (MBA thesis, Tulane University, June 1978); Stanford Research Institute, "Some Major Impacts of the National Space Program: V. Economic Impacts," N68-34387 (June 1968); Rockwell International, "Impact of the Space Shuttle Program on the California Economy."

55. S. Doctors, *The Role of Federal Agencies in Technology Transfer* (Cambridge: MIT Press, 1968); R. Lesher and G. Howick, "Assessing Technology Transfer," NASA SP-5067 (Washington, D.C. NASA, 1966).

56. H. Nieburg, *In the Name of Science* (New York: Quadrangle, 1966); E. Kennan and E. Harvey, *Mission to the Moon: A Critical Examination of NASA and the Space Program* (New York: Morrow, 1969); O. Wilkes and N. Gleditsch, "Optical Satellite Tracking," *Journal of Peace Research* 15 (1978): 205–25.

57. See E. Ginzberg, J. Kuhn, et al., *Economic Impact of Large Public Programs: The NASA Story* (Salt Lake City: Olympia Press, 1976); I. Mitroff, *The Subjective Side of Science: A Philosophical Inquiry into the Psychology of the Apollo Moon Scientists* (New York: Elsevier, 1974).

58. R. Bauer, *Second-Order Consequences* (Cambridge: MIT Press, 1969); B. Mazlish, ed., *The Railroad and the Space Program: An Exploration in Historic Analogy* (Cambridge: MIT Press, 1965).

59. Viz., chapter 16 in this book, and D. Goehuis, "The Changing Legal Regime of Air and Outer Space" *International and Comparative Law Quarterly* 27 (July 1978): 576–95; W. Schauer, "Outer Space: The Boundless Commons?" *Journal of International Affairs* 31 (Spring 1977): 67–80; J. Freeman, "Toward the Free Flow of Information: Direct Television Broadcasting via Satellite," *Journal of International Law and Economics* 13 (1979): 329–66.

60. Cass Schichtle, *The National Space Program: From the Fifties into the Eighties* (Washington, D.C.: U.S. National Defense University, 1983); W. N. Blanchard, *Evolution of Air Force Space Mission Command, Control and Communications:*

Proceedings of the Symposium on Military Space Communications and Operations Held At the USAF Academy, Colorado, on August 2–4, 1983, #A135021 (Springfield, Va.: NTIS); David Lupton, "Space Doctrines," *Strategic Review* 11 (Fall 1983): 36–47. Viz., J. Marriott, "Satellites," *Army Quarterly* 107 (July 1977) 291–97; R. Hansen, "Freedom of Passage in the High Seas of Space," *Atlantic Community Quarterly* 16 (Fall 1978): 345–57; W. Evans, "Impact of Technology on U.S. Deterrent Forces," *Strategic Review* 4 (Summer 1976): 40–47.

61. Schulman, "Nonincremental Policy Making." For other studies of space policy organizations, see Robert Richardson, "Technology, Bureaucracy, and Defense: The Prospects of the U.S. High Frontier Program," *Journal of Social, Political, and Economic Studies* 8 (Fall 1983): 293–99; Clayton Koppes, "The Militarization of the American Space Program: An Historical Perspective," *Virginia Quarterly Review* 60 (Winter 1983): 1–20; U.S. Office of Technology Assessment, *Civilian Space Policy and Applications* (Washington, D.C.: Government Printing Office, 1982).

62. W. Bainbridge, *Spaceflight Revolution: A Sociological Study* (New York: Wiley, 1976).

63. For example, see my other chapter in this volume.

64. Viz., U.S. Senate, Commerce, Science and Transportation Committee, *Hearings: Policy and Legal Issues Involved in Commercialization of Space* (23 September 1983); G. O'Neill, "Low Road to Space Manufacturing," *Astronautics and Aeronautics* 16 (March 1978): 24–32; M. Bartos, "Building Starscrapers in Orbit," *Civil Engineering* 49 (July 1979): 66–70; B. O'Leary, "Mining the Apollo and Amor Asteroids," *Science,* 22 July 1977, pp. 363–66; J. von Puttkamer, "Industrialization of Space," *Futurist* 13 (June 1979): 192–99; P. Collins, "Making It in Space," *Futures* 11 (October 1979): 439–41; M. Hopkins, *The Economics of Strikes and Revolts During Early Space Colonization,* P-6324 (Santa Monica: RAND, 1979); R. Salkeld, "Towards Men Permanently in Space," *Astronautics and Aeronautics* 17 (October 1979): 60–69.

65. Viz., A. Lacy, "Manufacturing in Outer Space," *Quarterly Review of Economics and Business* 18 (Spring 1978): 7–18; Various, "Enterprise on Enterprise," *Contemporary Business* 7 (1978): 1–189; R. Herendeen et al., "Energy Analysis of the Solar Power Satellite," *Science,* 3 August 1979, pp. 451–54; P. Glazer, "First Steps to the Solar Power Satellite," *IEEE Spectrum* 16 (May 1979): 52–58; P. Glazer, "Economic and Environmental Costs of Satellite Solar Power," *Mechanical Engineering* 100 (January 1978): 32–37; J. Glazer, "Domicile and Industry in Outer Space," *Columbia Journal of Transnational Law* 17 (1978): 67–118; U.S. House of Representatives, Science and Technology Committee, *Hearings: The Space Industrialization Act of 1979* (Washington, D.C.: Government Printing Office, 1979).

66. *Wall Street Journal,* 27 January 1984, p. 2; U.S. House of Representatives, Science and Technology Committee, "Space Commercialization" (report; October 1983); U.S. House of Representatives, Science and Technology Committee, "Commercialization of Land and Weather Satellites" (report; June 1983); *Business Week,* 20 June 1983, pp. 50–53.

67. R. L. Helmreich, "Applying Psychology in Outer Space: Unfulfilled Promises Revisited," *American Psychologist* 38 (April 1983): 445–50; M. Mechanic, "Planning and Design Frameworks for Space Colonies," *Ekistics* 50 (July/August 1983): 295–300.

Part I

Military Activities in Space

2

Space and the Preservation of Freedom

Harrison H. Schmitt

I delivered the following letter to President Reagan in the presence of Vice President George Bush and Counsellor to the President Edwin Meese:

17 August 1982

The Honorable Ronald W. Reagan
The White House
Washington, D.C. 20500

Dear Mr. President:

Events are taking place in space which will alter the course of history. Unfortunately, if we continue on our present path of relative indifference, it will be the Soviet Union that sets the course, not the United States of America.

Four hundred years ago in 1587, England, under the great Queen Elizabeth, was faced with a choice. The choice was to either take control of her destiny on the oceans of the Earth or to abandon that control to the forces of oppression.

The despot Philip of Spain was building an Armada of inferior technology and command, but of superior understanding of the role of sea power in the future course of human history.

Fortunately for us, and the free nation we were to become, Elizabeth made the right choice. In 1588, the superior technology and command (and a little luck) of a rebuilt English navy decisively defeated the Armada and took control of the commerce and the freedom of the Earth's oceans.

For some 300 years, the seed of individual freedom was protected, not without difficulty, but nonetheless protected by the shield of British dominance on the oceans. With the construction of the Panama Canal and World War II, we, the melting pot of millions who longed to be free, inherited this awesome task.

Mr. President, there is a new "ocean" in space and there is a new "Armada" being constructed by the forces of oppression. The Soviet Union's leaders long ago recognized that dominance in this new ocean of space will mean control of the destiny of peoples on Earth.

Through inferior technology but superior perspective, the Soviets are rapidly gaining the high ground of space, not just militarily, but of greater importance, they are gaining control of the beneficial "resources" of space, the weightlessness, the high vacuum, the unique view of the Earth, Sun and stars.

Mr. President, I believe that time and circumstances have placed you in the same historical position as Elizabeth four hundred years ago. Like Elizabeth, I also believe that you have no choice but to commit this nation irrevocably to humanitarian, commercial and military dominance in this new ocean.

If you do not make such a commitment, Mr. President, if *we* do not make such a commitment, then the light of freedom may flicker and die. The successors of Philip will have triumphed.

Mr. President, I am committed to work with your Administration on this matter in any way you see fit; however, at present those with responsibilities for the Nation's space activities are sorely in need of your guidance and inspiration.

Back in 1975, Neil Armstrong and I found that we had come to the same conclusions on the answer to the question: "What does it take to make America do great things?" We concluded that it takes four basic conditions: a base of technology from which we can reach to do the things to be done; a sustained period of national uneasiness; a catalytic event or situation which forces national attention on the directions to proceed; and, most critically, an articulate national leader who recognizes the great things we must do.

Mr. President, I believe all these conditions are met by our current situation and that you, sir, are that leader. The great thing to be done is to commit Americans to their destiny in space: the preservation of freedom.

I can guarantee you that a generation of young Americans awaits your call.

Sincerely,

Harrison H. Schmitt
U.S. Senator

The ensuing discussion with the president and the others showed they have a good awareness of our space activities, a strong perception of the excitement of space, and a commitment to current levels of activity, but the sense of historical perspective or the threatened freedom-lover's sense of urgency that was present 400 years ago is currently absent. Many other problems were obviously on their minds; the president's tax increase was pending in the Senate, and I was on record as opposing this bill. However, even as we search for solutions, we cannot let the problems of the present overwhelm our perspective of the future.

Today there exists a new ocean and a new challenge for mankind. I am referring, of course, to the limitless ocean of space. The nation on Earth which effectively utilizes technology to exploit the economic and military advantages of space will dominate activities on this planet at least well into the next century. In a military sense, space is the new "high ground" as well as the new ocean.

Until now, space and activities in space have been largely an adventure similar to the early voyages of Henry, Magellan, and Drake around the continents. Now, as did our ancestors at the shores of the seas, we must begin to grasp the opportunities of the frontiers of space.

We are already heavily dependent on space assets for our military and civilian communications system. The military also relies on space for reconnaissance, navigation, and meteorological data. That dependence is increasing each day. Industry, particularly the biomedical industry, is looking to space for new or more efficient manufacturing techniques. The lives of every American and every individual in the world are now affected directly or indirectly by activities in space.

The United States also has the technology to exploit the advantages of space in the defense of the United States and freedom. The questions are whether we will utilize our technology for our own well-being; whether we have the will to embark on a new indefinite adventure into this new ocean.

In the sixteenth century, the competition for power and influence on and around the seas was among nations that shared many political, economic, and cultural values. Today the competition for domination of space is between two competing political and economic systems. It is a matter of whether the United States and Western political and economic values of freedom and individuality will prevail or whether the totalitarian system of the Soviet Union will control the future of mankind. In that sense, the stakes are much higher than they were in the sixteenth century.

The economy of the Soviet Union is controlled by a military-industrial complex that has lost all control of its own system. It is part of a society based on military control and oppression of its peoples. It is a society whose economy requires continuous build-up of its military might and continuous expansion of its control of the world's resources. Retreat would mean the demise of its economy and the eventual loss of totalitarian control.

There is no doubt that the totalitarian system of the Soviet Union is dedicated to the destruction of freedom and Western values. During the period of so-called detente, the United States stopped expanding its military arsenal and began expanding trade and cultural relations with the Soviet Union. The

Soviets, however, continued to build up military might to the point well beyond anything that would be purely for defensive purposes.

The military-industrial complex in the Soviet Union cannot be turned off even by its own leaders. It is a system that employs most of its workers in the extracting of raw materials, in the refining of those materials, in the manufacture of armaments, and in the use of weapons in the field. To the degree that it can, the rest of the Soviet economy, particularly its agricultural sector, is struggling to support this military-based effort. To the degree that it cannot, ironically, that effort is supported by the food, technology, and capital of the West.

While the West has closely watched Soviet military expansion throughout the globe, little attention has been focused on Soviet space activities. During the last decade, the Soviet Union has pursued an aggressive space program that is almost purely military and political in nature. It has outpaced the United States in the number of launches per year as well as the variety of activities pursued. For example, in 1980 the United States had 13 space launches to the Soviets' 109. U.S. launches included commercial satellites as well as scientific missions. Virtually all Soviet launches were military in nature.

The United States has only recently conducted its first test of an antisatellite (ASAT) system. The Soviet Union has had a limited operational ASAT capability for almost a decade. The Soviets have maintained a nearly permanent manned space presence for years and now appear to be embarking on a program to build a large space station, capable of housing many individuals and activities, while we are still debating whether to even pursue such a venture for peaceful purposes, much less for the purposes of self-preservation.

The challenge before us today is to decide that the United States *will* embark on a program to sail the ocean of space, to develop its economic potential, and to protect our assets and interests in space and on the earth. Just as Britain dominated world politics for four centuries because of its technology, the United States could dominate because of its technological lead in space if only we have the will to do so.

To do this, three important requirements must be met to counter the current challenges of the Soviets:

1. The private and public infrastructure that supports this country's leadership in areas of science and technology must be greatly redeveloped and expanded. Scientists, engineers, and technicians, as well as engineering and manufacturing capabilities, are undercapitalized, underemphasized, and, for the most part, inadequate to the demands that we must place upon them. I emphasize that this rebuilding of our science and technology base must be as strongly supported in the civilian as in the military arena.
2. A national policy must be adopted and a commitment must be made for going into space and staying there permanently. This must be forthcoming

for our national security needs and for the exploration of the unique environments of space. The well-being of our own citizens as well as that of others throughout the world is at stake.

3. A goal of permanent presence of Americans in space, which is set in the context of the national policy, must be adopted and the first step must be taken very soon. In conjunction with that goal and its vigorous pursuit, we must continue to lay the research and technology base necessary for whatever future objectives and goals we may have to set. Our future policy options for solutions to most problems are limited only by an inadequate science and technology base from which to build.

If we do not address these requirements in responding to the challenge before us, we will see the psychological as well as the practical basis for a great deal of our scientific, engineering, and vocational education decline. European and Japanese efforts to capture aviation and aerospace markets not only will increase but will be increasingly successful; leadership in science and technology will leave our shores, particularly in science and technology related to space. Then the Soviets will capture the high ground of space by default. All of this adds up to a new Sputnik, a multinational Sputnik of real and frightening proportions. It is my prayer that the Reagan administration and future administrations recognize these imperatives.

Federal leadership in science and technology must come from the executive branch of government. The Congress lacks the confidence to make any major policy decision on such matters not only because of its lack of members with sufficient background and experience but because it must have an administration willing to carry out policy.

When I look at the excitement among the American people and the people of the world that has been generated by space and space technology, I remain optimistic that we will meet the challenge that is before us. It will, however, require shaking up the leadership in government that until now has failed to see either the challenge or the benefits that space offers this nation and humankind.

The issue before us is freedom. There is nothing more precious. We have offered our lives to protect it. The challenge of space is really the battle to preserve freedom for future generations as well as to spread freedom to more parts of the world today. We must take that challenge.

3

Space Militarization: A Costly Mistake

John Joseph Moakley

The United States is at a turning point on the crucial issue of space weapons and national space policy.

President John F. Kennedy set the tone for the U.S. space effort over two decades ago: "For the eyes of the world now look into space, to the moon, and to the planets beyond, and we have vowed that we shall not see it governed by a hostile flag of conquest, but by a banner of peace and freedom. We have vowed that we shall not see space filled with weapons of mass destruction, but with instruments of knowledge and understanding." It was this brand of visionary idealism that set the stage for the spectacular successes of the Apollo moon landings, the planetary probes, and even the Space Shuttle. We were not going into space solely for our own advantage but for the benefit of all humankind.

On 23 March 1983, President Ronald Reagan alluded to a new vision of space's potential. Where Kennedy called on us to put a man on the moon, Reagan has challenged us to produce a new, futuristic, space-based missile defense system. It seems apparent that the optimism and universal goodwill which engendered the early space program have given way to a new determination to exploit space of military advantage.

Since Reagan's speech in March of 1983, there has been much discussion concerning the feasibility and desirability of developing a space-based ballistic missile defense system (BMD). Proponents of such a system argue that it would provide the United States with an alternative to the current strategic faith in mutually assured destruction (MAD). Certainly, we must question the wisdom of using the threat of total annihilation to maintain peace, but I do not believe that we will find an alternative to MAD in a space-based BMD system. Aside from the many mechanical and technical drawbacks of developing such a system, there would be tremendous costs. Further, it is pure folly to believe that the Soviet Union will watch calmly while we develop a new grandiose war-fighting capability. Instead, it will match us weapon for weapon, dollar for dollar. Deploying such weapons in space will most likely result in not only a matching Soviet "defensive" effort in space but an unlimited offensive nuclear arms race by both sides to overcome the other's defensive deployments. It seems to me that it is far more sensible to seek an alternative to MAD through the pursuit of bilateral and verifiable arms control and arms reduction agreements.

Although much of the so-called star wars technology being discussed is still in the planning stage, the weapons race in space is at a critical point with the development of antisatellite weapons (A-SATs). The United States first deployed an A-SAT system in the early 1960s. That system was retired from service in 1975. The Soviets have been testing a crude A-SAT since 1968; it is ungainly and unreliable. Launched from atop a huge SS-9 rocket booster, it has achieved only a 50 percent success rate to date. Most importantly, the Soviet system does not have the range to threaten the most crucial U.S. military satellites.

Unfortunately, the race is continuing, and the Reagan administration is about to usher in a new era in antisatellite weaponry as we begin flight testing a new, sophisticated A-SAT. The U.S. A-SAT will be clearly superior to its Soviet counterpart. It is small enough to be carried aloft by high-flying F-15 fighter aircraft. Once fired from the F-15, the missile is propelled by a two-stage rocket. The intended target is actually destroyed by a 12-by-13-inch canister, called a miniature homing vehicle, that simply rams the satellite at high speed. Because of the A-SAT's small size, it will be extremely difficult to verify its deployment, thus making a treaty to ban such weapons unlikely. Every F-15 will be a potential A-SAT platform in Soviet eyes. In contrast to the Soviet system, the U.S. weapon could be capable of destroying crucial Soviet early warning and communication satellites.

It is important to note that the military forces of both the United States and the Soviet Union have long relied on satellites for a variety of peaceful support functions such as weather, navigation, early warning, treaty verification, and reconnaissance. These functions are used to enhance our overall security and maintain peace. Satellite reconnaissance, for example, allows the superpowers to monitor each other's military activities. Placing weapons in space that might threaten these satellites will raise rather than lower the chances of a devastating nuclear war on Earth. With the central nervous systems of the immense superpower war machines already in orbit, just the existence—let alone the use—of these weapons will turn every computer malfunction and unexplained mechanical failure into a pretext for war.

Thus, we are at an important threshold with regard to space weapons. We can either make a good-faith effort to negotiate a ban on such weapons or we can destine this and future generations to a new round of arms competition. Beyond the strategic problems, there is the issue of cost. We all know only too well the ill effects the current arms race has had on our economy and our social programs. Our cities are crumbling, we cannot afford adequate health care for our citizens, and our young people cannot afford a decent education. I truly believe that if we embark on an extravagant arms race in space—a race that will threaten, not guarantee, our security—we will pay a grave price in terms of the overall quality of life for our people.

Original projections placed the total cost of the U.S. A-SAT program at $1.7 billion. Soon that figure grew to $3.6 billion. There is a General Accounting Office report which concludes it will be even more, probably in the "tens of billions of dollars." This figure does not begin to reflect the price of more grandiose space weapons currently being contemplated. Cost estimates for a space-based BMD system run into the hundreds of billions, even trillions, of dollars.

President Eisenhower, a Republican, in warning about the dangers of the continuing arms race, said:

> The worst to be feared and the best to be expected can be simply stated. The worst is atomic war. The best would be this: A life of perpetual fear and tension, a burden of arms draining the wealth and labors of all peoples, a wasting of strength that defies the American system or the Soviet system or any system to achieve a true abundance of happiness for the peoples of this Earth. Every gun that is made, every warship launched, every rocket fired, signifies in the final sense a theft from those who hunger and are not fed, those who are cold and are not clothed.

Let us realize that an arms race in space is a dead-end street that, at best, will leave us warped with fear. And let us realize now, before it is too late, that the cost of such a race will devastate our economy. National security means more than the number of weapons we possess. It means, as well, the vitality of our economy and the quality of life we offer all our citizens.

The Congress should soundly reject the Reagan administration's dangerous "star wars" proposal. I urge the nation to adopt a policy with the following components:

1. A commitment to a moratorium on flight tests of the U.S. antisatellite weapon against objects in space, as long as the Soviet Union maintains its current moratorium on further tests of its own system.
2. A commitment to reopen formal negotiations with the Soviets on proposals to limit or ban the deployment of A-SAT weapons.
3. A reaffirmation of U.S. commitment to the principles embodied in the Antiballistic Missile (ABM) Treaty.

4. A commitment to refrain from tests that would require de facto abrogation of the ABM Treaty, provided the Soviet Union also refrains from any such tests.
5. A commitment to broad-based negotiations with the Soviet Union designed to reach a mutual and verifiable agreement banning all weapons from space, and to enhance U.S.-Soviet cooperation in the peaceful uses of space.
6. A commitment to a vigorous research and development program on space technology as a hedge against any possible Soviet breakout or abrogation of the ABM or other relevant treaties.

I urge that all those Americans who are concerned with arms control move this issue up on their list of priorities. This generation is the first to look to the heavens and know the stars are within reach. Shall our children pursue this new destiny peacefully, in the spirit of exploration? Or shall they view outer space as yet another arena for the futile attempts of one nation to gain temporary military advantage over another? The choice is before us now. Let us work to keep space free from weapons.

4

Space Militarization and the Maintenance of Deterrence

Thomas Blau and Daniel Gouré

The strategic policy of the United States remains committed to deterring nuclear war through the threat of massive punitive retaliation. Yet, the enormous nuclear arsenals of the Soviet Union increasingly place the United States at risk and reduce the credibility of the retaliatory threat. The potential use of U.S. strategic power is limited to only the most extreme circumstances (e.g. defense against a Soviet attack) because of concerns for escalation control, worried allies, and growing Soviet strategic (and theater) capabilities.

Relations between the United States and the USSR are defined mainly in military terms. Despite arms control and other areas of coincidental national interest, the active use of military force for political leverage and in defense of national interest by both the U.S. and USSR shows no sign of decreasing. Still, preventing direct conflict, particularly in nuclear war, will continue to be central to superpower relations and to the policies of the two blocs. However, avoidance of nuclear war does not negate the Soviets' attempt to exploit politically their military capabilities. Thus, a military "balance" that both favors the Soviet Union and inhibits U.S. flexibility and initiative conflicts with the global character of U.S. national interests.

It is not surprising that the United States is profoundly uncertain about the use of its military power. Driving the uncertainty is a growing sense of national and personal insecurity regarding U.S. inability to achieve a stable strategic balance. Such a balance was pursued in the 1970s when the United States acquiesced in the Soviet drive for strategic equality. At that point, the Soviets were to have accepted a roughly equal deterrence and to have joined the United States in a search for additional stability in the superpower relationship. This has not happened.

Both the Reagan administration and the antinuclear freeze movement see the present situation as dangerous and potentially unstable. Both attribute this to the same factors: the growth of nuclear arsenals, the hair-trigger nature of modern command and control systems, and the inadequacy of current nuclear strategies. Yet, they differ radically as to the proper prescription for action. The main responses have on the one side been increasingly complex and hard to understand (such as the late "dense pack" plan for basing new U.S. intercontinental ballistic missiles) and on the other simplistic (such as unilateral disarmament). The Soviet Union, however, suffers less from these problems, a reflection, in part, of Moscow's having lived for more than 20 years with the spector of annihilation. For the U.S. the problem is of recent origins (approximately ten years).

Deterrence Stability

The U.S. deterrent is in many key respects unchanged since the mid-1960s. It now faces vastly more threatening forces on the other side. Until recently, suggestions for restructuring or upgrading U.S. strategic forces focused almost completely on strategic offensive forces. In part, this reflected the dominance of offensive thinking and of coercive deterrence embodied in the doctrine of mutual assured destruction (MAD), the latter viewing defenses as ineffective if not actually destabilizing. This view was codified in the 1972 Antiballistic Missile (ABM) Treaty, the centerpiece of the Strategic Arms Limitations Talks (SALT). This treaty not only banned existing strategic defensive systems but also effectively prohibited deployment of future, more effective ones.

U.S. strategic forces and concepts, to date, have been shaped by marginal improvements to existing systems and by the effort to draw the Soviets into arms controls measures. There has not been much effort to respond with fundamentally new or different U.S. capabilities or to address the broad, long-term viability of the U.S. strategic deterrent. The problem is that Soviet counterforce capabilities imply a destabilizing interest in a first-strike. The theory of arms control and our confidence in stability are thereby called into question.

Some, such as key figures in the Reagan administration, do not believe deterrence is automatic and inevitable in the new strategic balance. They believe that there is a need to negate additional Soviet advantage to close what the president calls the "window of vulnerability." But simply adding to U.S. offensive forces would at some point increase the frustration and cost of trying to negate the massive Soviet build-up, and raise a new moral question of peacekeeping through the threat of mass annihilation.

"Star Wars"

It is understandable, then, that the Reagan administration would try to rethink U.S. strategy, as suggested by the president's "star wars" speech on

23 March 1983. The administration, like Carter's before it, has recognized the potential military applications of outer space. As a result, it is giving special attention to the merits of deploying a space-based strategic defense of U.S. territory. The administration plans an accelerated research program over the next five years, costing about $1 billion a year, to give the United States this capability.

Using outer space for strategic defense of the United States could eventually shift the military, political, and moral basis of deterrence. More broadly, the idea of building defenses against nuclear weapons, while not new, again raises old issues: What are the national security objectives of the United States? How should U.S. forces be configured and employed in the attainment of those objectives? How stable will such a future world be?

The Reagan administration emphasizes the primacy of deterrence of the USSR. Others, such as arms control experts Herbert Scoville, Richard Garwin, and Paul Warnke emphasize the primacy of the strategic arms negotiations. They tend to reject options that involve newer, larger, or more elaborate U.S. weaponry, that mitigate U.S. vulnerability to a Soviet nuclear attack, or that can evoke Soviet opposition which would then damage the atmosphere of arms control negotiations. They believe the path to peace lies in MAD, which they believe equally compels both sides to avoid war. Underlying the disagreement between the two sides on U.S. policy is a basic assumption regarding the primacy of vague national interests and concerns versus shared, transcendent values.

The administration interest in outer space lies in maintaining the (now questioned) strategic balance, in investigating space for military applications, and in reducing the considerable lead taken by the Soviets in using space for military purposes. Even without futuristic developments for strategic defense of U.S. territory, outer space is now critical to military command, control, and communication (C^3), warning, reconnaissance, and arms control verification, among other functions. If one side's satellites devoted to these tasks were destroyed, it could be rendered "blind, deaf, and mute." It is, therefore, especially sobering that the Soviets have conducted a score of satellite-killer (A-SAT) tests; the United States, one.

The USSR could conceivably combine military leadership both in advanced space applications and traditional military capabilities, thus producing grave political problems (at the very least). Space forces, for example, might offer an aggressor the chance to control the scale and scope of conflict, thereby greatly reducing the cost of aggression. For the defender (i.e. the United States), however, space may be a critical dimension in frustrating any Soviet planning for central war.

Although some in the Reagan administration talk of developing practically perfect strategic defenses in the (far off and expensive) future, *modest* gains to strategic defense could be very important *now*. Even a limited ability to defend the United States or its military assets could provide disproportionately large deterrent benefits, if only by complicating all the decisions made by an aggressor in planning the attack. Because of the enormous burden of achieving coordination faced by a nuclear attacker, confidence reduction—especially in the early moments, when the attacker must deal with warning and C^3 assets—may well critically blunt the edge of a potential first strike and thereby deter it, thus peace would be promoted.

Rethinking Required

If U.S. military applications of space can undermine the utility of the potential aggressor's superior forces, then there is reason to revise certain attitudes about space. These attitudes are:

1. that space is sanctified for peaceful scientific and humanistic development, and that military activity is not precedented, desirable or feasible and is subject to being banned in any event;
2. that space can only passively supplement routine, terrestrial political-military concerns (e.g. arms control verification); and
3. that in the extreme, space can be no more than an arena of localized military operations between A-SATs, target satellites, and perhaps defensive satellites (DSATs).

To assume that space is militarily useless to the U.S. can turn out to be a self-fulfilling prophecy if inaction permits the USSR should establish a visible breakthrough. The impact then may be all the greater because of the gulf between the initial passive, skeptical view of the military use of space and the results exploited by the other side. Surprise will result in shock (a central concept in Soviet military planning).

Following are potential military uses of space over the current, medium, and long terms:

1. There is an immediate need to support Earth-based activities and military forces from space, including early warning of attack, C^3, navigation, intelligence gathering, weather monitoring, and even direct U.S.-Soviet communications via the "hotline." Most of these functions relate not to nuclear war but to the day-to-day, peacetime operation of U.S. military forces, the verification of arms control agreements, and the surveillance of global military activities.
2. The prospect of combat in space is raised by the impending deployment of workable antisatellite weapons, which will threaten many current support satellites. This is a medium-term issue.

3. Space might also be utilized to defend the United States and its allies from the growing threat posed by Soviet strategic and theater-nuclear forces. In the long term—after the year 2000—a space-based defense against missile attack is possible.

The United States, its allies, and the Soviet Union are now critically dependent on space for the performance of a large number of national security functions. Space-based sensor systems provide early warning of ballistic missile attacks. To provide global strategic and tactical communications, the United States maintains the Defense Satellite Communications System (DSCS), the Fleet and Air Force Satellite Communications Systems (FLATSATCOM and AFSATCOM, respectively), NATO Communications Satellite System, and the Satellite Data System (SDS). The Soviet Molniya system provides Moscow with a similar capability.

Both the United States and the Soviet Union depend on a variety of satellite-based sensors for purposes of surveillance, reconnaissance, and strategic targeting. While the United States is believed to be ahead in high-resolution photographic and image-sensing systems, the Soviet Union has several unique systems, in particular, its radar and electronic oceans reconnaisance systems (RORSAT and EORSAT, respectively). It has recently been reported that the Soviet RORSAT system is capable of providing "real-time targeting" of U.S. warships for Soviet cruise missiles. This means that the Soviets could engage U.S. ships at extremely long range (up to 300 kilometers), outside their air-defense envelope and without requiring a nearby ship or plane to provide in-flight target tracking.

In addition, there are meteorological and navigational systems that provide information necessary both for accurate strategic targeting and for the operation of conventional and naval forces. The U.S. NAVSTAR Global Positioning System provides accurate (within 30 meters) time-of-arrival guidance to any point on Earth. The Soviets intend to deploy a similar system.

Satellite surveillance systems help maintain superpower peace, verify arms control agreements, and guard against nuclear proliferation. In 1978, President Carter noted that "national technical means of verification," central to enforcement of the strategic arms control agreements, relied heavily upon satellite systems. Their ability to provide reliable data on a potential adversary's state of military preparedness has allowed the superpowers to limit their own military plans accordingly.

U.S. Space Capabilities

Central to the further exploitation of space, whether for military or civil purposes, has been the ability to place and to maintain men and machines in space. The U.S. Space Shuttle makes it possible to place 65,000 pounds of payload in low Earth orbit (1,000 miles above Earth). This will enable the United States to orbit larger, more complex satellites. Additionally, the Shuttle could be used to support construction of a permanent station in space. The Soviet Union appears to be on the verge of creating just such a permanent station and are building their own fleet of shuttles.

Space-based systems have for many years served as a force multiplier, enhancing the effectiveness of more conventional military force. However, it is the likely deployment in space of active weapons, as distinct from the current passive systems, that promises to alter radically the nature of war and military planning.

Currently, there are no weapons in space, and their future deployment may be limited by a series of treaties. The 1967 Outer Space Treaty bans all nuclear weapons in space, on the moon, and on other celestial bodies. The 1972 ABM Treaty also bans the deployment in space of systems capable of attacking ballistic missiles. The United Nations has proclaimed space to be a "zone of peace." No treaty, though, prohibits the passage of nuclear-armed missiles through space. Nor does any treaty directly prohibit the deployment of ground-based antisatellite weapons systems, such as those currently possessed by the USSR and under development by the United States. While some have argued that international law views space as a place where war is banned, this zone is increasingly indispensable to terrestrial military planning and force posturing.

U.S satellites are increasingly threatened by the Soviet A-SAT system, the only one now operational. The Pentagon recently examined its utility during a nationwide military exercise simulating a Soviet nuclear strike on U.S. forces. (Also simulated was a Soviet ballistic missile defense (BMD) against U.S. retaliation forces.) The Soviet A-SAT is a large conventional or nuclear-armed satellite launched by a version of the SS-9 ICBM into a co-orbital intercept with the target satellite. The Soviet system is effective only against satellites in low Earth orbit, although use of a larger booster could allow the system, particularly if armed with a nuclear warhead, to attack U.S. communications and early warning systems in geosynchronous orbit (22,300 miles) by means of electromagnetic pulse. Such a capability has long been sought by the Soviet Union to give it the means for successful nuclear warfighting.

The United States is taking steps to improve the survivability of its satellite systems against attacks in space or on the ground. An announced $18 billion satellite-survivability program will provide protection for vulnerable Earth stations and communications facilities as well as for the satellites themselves. Some systems, such as defense communications satellites, have "silent spares" already in orbit that can be used to replace operational systems. Other avenues of protection include rapid reconstitution capability, hardening of satellites against shock or nuclear effects, and enhanced ability to maneuver out of harm's way. In addition, the United States is currently developing a comprehensive threat management and surveillance system to track all U.S. satellites and to warn of an impending attack.

Moreover, the United States has sought to develop a countervailing threat to Soviet A-SATs. The U.S. system will consist of a miniature homing vehicle mounted on a two-stage rocket and carried on an F-15 or similar aircraft. Launched in the air, it will home in on Soviet satellites and destroy them by direct impact. Although the U.S. system will not reach the high altitudes of the larger Soviet rockets, its flexible launch mode will permit its use against low-orbit satellites from virtually any spot on the globe.

The growth in the military and civil uses of space has created a gray area of systems and functions having both military and nonmilitary application. The U.S. Space Shuttle transports military as well as civilian payloads, and the Soviets have repeatedly insisted that the Shuttle can be used to "capture" their satellites. Space stations can likewise be used for a variety of purposes. The Soviet Salyut space station may well be intended as a military observation post and a base for Soviet A-SAT weapons.

These early generation space systems are the forerunners of a broad array of potential support and active military capabilities in space. Advanced infrared technology and mosaic-array sensors may provide the basis for detecting aircraft, cruise missile, and ballistic missile threats from space. One such system, currently being tested under the name Teal Ruby, may serve as the space-based sensor component of a large air defense capability. Advanced concepts in BMD include use of a mosaic-array sensor overlay which, when launched into the path of an incoming ballistic missile attack, discriminates targets for a ground-based antiballistic missile defense. Other proposed U.S. systems include laser satellites for communicating with submerged submarines; radar satellites, similar to one now operated by the Soviet Union, for ocean surveillance; and improved photographic and sensor satellites for intelligence gathering, submarine tracking, and arms control verification. Coupled to improved delivery systems and "smart" munitions, they could make the conventional battlefield of the future as deadly as the present nuclear battlefield.

Breaking Out of the Future

Even more revolutionary is the prospect of placing BMD as well as A-SAT weapons in space. If a defensive system could be put into space, say proponents, ballistic missiles might be intercepted in the boost phase, prior to separation of the warheads from the missile itself, thus simplifying the defensive problem.

In 1972, it was argued that banning BMD made sense because the technology was primitive and unproven and because the attacker could always develop means to counter any defense. Since both sides were believed to be focusing on deterrence by means of a capability for assured destruction of "soft" targets (i.e. cities), additional weapons would serve no purpose; they would be "overkill." Defenses for cities, it was argued, would have to be 100 percent effective to preclude catastrophic damage, so no defense appeared feasible. However, the Soviets have continued vigorous ABM efforts arguing that steps to protect their nation are essential to the credibility of Soviet defense policy. Furthermore, the possibility of placing a BMD system in space, either alone or in concert with other defenses in a layered effect, also invites re-examination of the argument for writing off strategic defense, especially if one's goal is less ambitious than guaranteeing the safety of cities; namely, to preserve some U.S. ICBM in the event of a Soviet first strike. It is somewhat paradoxical that many opponents of strategic defense are also critics of recent efforts by several U.S. administration to develop limited nuclear options. One consequence of deployment of strategic defenses would be to deny either side the ability to execute militarily effective limited strikes. Massive retaliation would thereby be secured.

Ballistic missile defense (BMD), in or out of space, now draws interest for a number of reasons. First, even a marginally effective ability to increase an aggressor's uncertainty by complicating his military planning would buttress deterrence. Second, such defenses reduce the potential for catastrophe from an accidental launch, or from a small or third-party nuclear attack. It may be particularly effective against so-called limited nuclear options in which one would expect the Soviet Union to widen their deployment. If it could defend its strategic nuclear forces against attack, the United States would not necessarily have to respond to limited attacks by retaliation against the USSR. Also, strategic defenses against missiles reduce the plausibility of preemption and cheap victory—again, increasing stability.

Strategic defenses may make sense precisely because they can provide a plausible middle ground between unilateral disarmament, on the one hand, and a continual buildup of offensive forces, on the other. BMD could be the foundation for a panoply of national defense capabilities, including cruise missiles and submarine-launched missiles, for defense against a first strike. *A defensive posture could provide a practical method of reducing both the likelihood and the consequences of war.* There is good evidence that the Soviet Union is pursuing a strategy now with improved air defense, dual capable anti-air/anti-missile weapons and a renovated BMD capability. Washington recently accused Moscow of violating the ABM Treaty by constructing new ABM battle management radars.

A wide range of technologies are currently being investigated for use in space. Some involve the use of near-term technology in the form of homing vehicles, pellet charges, or other devices intended to be launched on warning of attack and designed to use the kinetic energy from direct impact with the target. One proposal that has received public attention is the High Frontier concept, whereby existing technologies would be used to deploy a fleet of more than 450 satellites armed with multiple homing rockets for a global BMD. This system is distinctive in that it combines available technology with space-basing. High Frontier advocates insist that their system can be deployed within a few years and at relatively low cost. Their critics disagree sharply.

Potentially far more dramatic would be the deployment of directed-energy weapons—lasers, charged particle beams, and X-rays—which can pass easily through empty space. A Congressional study has suggested that "the potential of high-energy laser technology for altering the strategic balance between the U.S. and the USSR is presently unique." Likewise, a report by the General Accounting Office has recommended that the Defense Department speed up the development and deployment of "a constellation of laser battle stations in space." An orbiting directed-energy BMD system could blunt the cutting edge of a Soviet strategic attack as well as defend U.S. satellites from Soviet A-SATs.

The United States may be able to deploy a rather simple, low-power laser weapon within ten years; it probably would not be able to serve in a ballistic missile defense role but could defend U.S. satellites against A-SAT attack. More sophisticated systems might be developed within fifteen or twenty years, able to attack not only ballistic missiles but aircraft and even low-flying cruise missiles. One proposal under study is to deploy X-ray lasers powered by a nuclear explosion. In the event of an attack such satellites would detonate themselves, sending out beams of X-rays at approaching missiles. Unlike laser satellites, such a system would not need to carry large fuel supplies.

The potential multiple applications of directed-energy technologies make them especially interesting. Such systems could also help protect U.S. warships and even cities against attacks from high altitude aircraft or cruise missiles. Clearly, positioning such weapons in space would give them a range and coverage capability denied to any ground-based weapon. Such technologies could also be employed against radars and other electronic systems.

No one has yet tried to deploy an array of armed satellites. Merely procuring and launching them would be a very costly and complicated enterprise. Estimates of the costs by critics go as high as $300 billion, the likely figures being in the $50 to $100 billion range. While these figures are very high, they are comparable to potential "traditional" responses to Soviet buildups, involving clearly threatening offensive weapons. Additionally, directed-energy weapons require extremely high pointing accuracy, exact target trackers, and very high power output if they are to be used against ballistic missiles. Such technologies are only in the development stage. Currently, the U.S. government is spending almost $500 million on researching directed-energy weapons, and if supporters have their way the sum will soon climb to several billion dollars.

The "awesome" potential (as a U.S. government study put it) of new space-based technologies is prompting a gradual readjustment of military doctrine and strategy with respect to space-based systems. In September 1982, the Air Force created a new unit—the Space Command—to exercise consolidated control over the vast array of Air Force space programs. Much as the development of the Strategic Air Command presaged a fundamental shift in U.S. strategy and doctrine, so the advent of Space Command suggests an attempt to coordinate U.S. operations in space.

The real innovation of President Reagan's March 1983 "star wars" speech lies not in his suggestion that a new ballistic missile defense might successfully repel attack but in his call to "break out of a future that relies solely upon offensive retaliation for our security." The president asked, "What if free people could live secure in the knowledge that their security did not rest upon the threat of instant U.S. retaliation to deter a Soviet attack; that we could intercept and destroy strategic ballistic missiles before they could reach our own soil or that of our allies?" Perhaps, then, the determining factor in the military exploitation of space by the United States will be a change in U.S. strategic policy and its objectives—most notably, a movement away from strategies of mutual assured destruction (embodied in SALT, and the Carter administration), or unilateral disarmament, and toward survivability.

The Soviet aim implied by A-SAT experiments and by Moscow's drive for superiority is to increase their certainty of success should they choose to attack. A minimum aim of U.S. strategy under pressure from the comprehensive Soviet buildup, then, should be to buttress the strategic deterrent by increasing the uncertainty of Soviet planners about the probable success—by any standard—of an attack on the United States. More important, however, is the role of space-based defenses in repairing the damage to U.S. strategic nuclear policy and commitments which has resulted from Moscow's acquisition of the capability to dominate conflict at the conventional, tactical nuclear, and strategic levels.

The opportunities for both military advantage and mutual stability arising from the exploitation of space suggest that the possibility of maintaining space as a "demilitarized" area has already been denied. The threat of hostile action against key U.S. satellite support systems is potentially great enough to mandate a response, if only to maintain the currently precarious strategic balance. More important, strategic defenses become increasingly attractive in light of the Soviet capability to employ force and to neutralize U.S. reaction at almost every potential level of conflict. Strategic defenses, of which space systems may be a part, seem to offer the prospect of moving away from the moral, political, and strategic precipice represented by assured destruction. The challenge will be not only to create new technology but to develop the domestic and international institutions and understandings which can manage the transition to a less hair-trigger world.

5

Space Militarization and International Law

Harry H. Almond, Jr.

The military activities in outer space have given rise to new expectations concerning the law. The formal sources of law associated with outer space are identified as the United Nations Charter, treaties and other international agreements relating to outer space, and customary international law that has gradually developed to regulate the activities of states in outer space. Resolutions of the United Nations General Assembly are said to have some "legal effect," depending upon the subject and the voting. These treaties and agreements include those concerned with arms control and disarmament, and if armed conflict should ever break out, we would also include the law of war, i.e. the law and the treaties relating to the conduct of hostilities.

The law governing the use of weapons in warfare has gradually separated into three categories. The first category, referred to under the Roman law notation of *jus ad bellum*, governs the initiation of threat or armed attacks among states. The present manifestation of such law appears in Article 2(4) of the United Nations Charter:

> All members [of the United Nations] shall refrain in their international relations from the threat or use of force against the territorial integrity or political independence of any state, or in any other manner inconsistent with the Purposes of the United Nations.

Article 2(4) is understood by many commentators to mean that states have renounced war as an instrument of policy; instead, according to Article 33, they are to settle their disputes by negotiation, enquiry, mediation, conciliation, arbitration, judicial settlement, resort to regional agencies or arrangements, or other peaceful means of their own choice.[1] Many of the proposals for arms control or for moderating the impact of conflicting policies among states are based upon the direction in Article 33, because it is believed that through reliable fact-finding the sources of the conflicts can be relieved.

A second category of law governs the conduct of warfare, and is identified in the Roman Law notation of *jus in bello*, i.e. the law of war. This law appears in a variety of treaties and agreements, including the well-known Geneva Protocol of 1925 that forbids the use of lethal chemical agents and incapacitating gases, and is emerging from the custom and usages of states in armed conflict.

A third category of law, which is of special interest in the context of space militarization, relates to the control of weapons. This form of international law appears in the contractual or treaty provisions identified as the arms control and disarmament agreements. Preeminent among the agreements is the United Nation's early arms limitation attempt, the Outer Space Treaty of 1967.

Three illustrations with respect to the earlier attempts to eliminate weapons in their entirety might be mentioned to illustrate the difficulty of forging meaningful arms control agreements. The first of these, the Versailles Treaty of 28 June 1919, provided for the total disarmament of Germany after its defeat in World War I. The following provisions suggest the tone of the disarmament features included in the treaty:

> (Article 159) The German military forces shall be demobilized and reduced as prescribed hereinafter.

> (Article 160) Preparation for war is forbidden. . . . The Great German General Staff. . . . shall be dissolved and may not be reconstituted in any form.

> (Article 170) Importation into Germany of arms, munitions, and war material of every kind shall be strictly prohibited.

> (Article 178) All measures of mobilization or appertaining to mobilization are forbidden.

The article of the Covenant of the League of Nations on disarmament, which entered into force on 10 January 1920 provides a further illustration of the attempts following World War I to deal with arms and arms stockpiles—with the expectation that if these could be controlled or eliminated, warfare itself would be eliminated. Article 8 reads:

> The Members of the League recognize that the maintenance of peace requires the reduction of national armaments to the lowest point consistent with national safety and the enforcement by common action of international obligations.

> The Council [of the League], taking account of the geographical situation and circumstances of each State, shall formulate plans for such reduction for the consideration and action of the several Governments. Such plans shall be subject to reconsideration and revision at least every ten years.

After these plans shall have been adopted by the several Governments, the limits of armaments therein fixed shall not be exceeded without the concurrence of the Council.

The Members of the League agree that the manufacture by private enterprise of munitions and implements of war is open to grave objections. The Council shall advise how the evil effects attendant upon such manufacture can be prevented, due regard being had to the necessities of those Members of the League which are not able to manufacture the munitions and implements of war necessary for their safety.

The Members of the League undertake to interchange full and frank information as to the scale of their armaments, their military, naval and air programmes and the condition of such of their industries as are adaptable to warlike purposes.

Germany denounced the Treaty of Versailles after little more than a decade. A decade after that saw the demise of the League of Nations, at the start of World War II. Numerous proposals offered prior to that time—notably the Washington Naval Agreements of 1922 and the Geneva Protocol of 1925—had limitations on weaponry, but none of these effectively reduced the use of arms during World War II. Neither the arms limitation attempts of the Versailles Treaty nor the League of Nations did much to impede (and may have inadvertently encouraged) an arms buildup.

A third path toward arms control has been pursued through the United Nations, where proposals and even treaties of various kinds have appeared in the four decades that followed the adoption of its charter. In addition, numerous resolutions with respect to controlling weapons and their use have been introduced in the United Nations General Assembly. While these do not have the force of law—at least for those who insist that law must be "legally binding" and fully enforceable—they do provide a basis for more effective, enforceable agreements. Among these are general agreements banning the testing of nuclear weapons in the atmosphere and in outer space, prohibiting the transfer of nuclear weapons and the technical data for producing them, and curbing the proliferation of nuclear weapons.

There are also disarmament and arms control efforts that have been directed to specific arenas. For example, the Latin American Nuclear Free Zone Treaty provides for a zone free of nuclear weapons in Latin America and the adjacent high seas, but it is marred by reserving the right of states in that area to pursue nuclear devices they deem to be peaceful. The Seabeds Disarmament Treaty is directed to disarmament on the beds of the high seas, but is weakened because it does not affect waters extending 12 nautical miles from the coast, and, with the Law of the Sea Treaty, its limits may eventually be extended, through state practice, to the high seas 200 nautical miles from shore.

Proposals specifically aimed at controlling arms in outer space may be included in the category of the general arms control agreements. Because the

U.N. treaty dealing with the military uses of outer space is both the most recent and most germane agreement on regulating these weapons, we will examine it in detail.

The Outer Space Treaty

The primary provision for arms control in outer space appears, at present, in Article IV of the Outer Space Treaty of 1967:

> State Parties to the Treaty undertake not to place in orbit around the Earth any objects carrying nuclear weapons or any other kinds of weapons on celestial bodies, or station such weapons in outer space in any other manner.
>
> The moon and other celestial bodies shall be used by all State Parties to the Treaty exclusively for peaceful purposes. The establishment of military bases, installations and fortifications, the testing of any type of weapons and the conduct of military maneuvers on celestial bodies shall be forbidden.

Article IV provides at best for only a partial demilitarization of outer space. It ignores, for example, prohibitions on the use of weapons in combat, and cannot and does not cover such uses in the event that the space powers engage in armed conflict. It leaves open the problem of weapons activities—or military activities in general—with respect to readiness and preparations relating to self defense. Other articles in the treaty, as construed by the United States members of the negotiating team, show that it was expected that all states, pursuant to the United Nations Charter, would reserve their right of self-defense. A further weakness in Article IV is that it does not preclude the right of states to provide for the self-defense of space stations, satellites, etc. Because self-defense of such objects naturally would entail military measures, the addition of military capabilities to space objects is, in fact, allowed under Article IV; self-defense from other objects in outer space, or from attacks launched from the atmosphere or the surface of the Earth, is permitted. It must be recognized that self-defense includes the ability to neutralize (i.e. counterattack) the sources of such threats. This, then, includes antisatellite weapons. Moreover, provisions in Article IV relating to military activities, military facilities, and military maneuvers are directed exclusively to *natural* bodies in space, and not to manufactured objects. Finally, it is not even clear whether asteroids are to be treated as "celestial bodies." These are all matters that must be clarified in the practice of states.

Article IV prohibits only nuclear and other weapons of mass destruction. The weapons to be included in this category have been the subject of controversy. The United States believes that a resolution adopted in 1948 in the United Nations Security Council adequately identified such weapons as nuclear (the term "atomic" was then used), chemical, biological, and radio-active. All other weapons are permitted in outer space—and, to that extent, are identified as conventional weapons. The increasingly destructive force of such weapons cannot be overlooked.

Moreover, the delivery systems for bringing weapons to bear in an attack are not regulated. Such systems for both conventional and mass destruction weapons would be identical. Therefore, the design development to solve problems of maneuver, control, target accuracy, and so on can be readily carried out without infringing the Outer Space Treaty obligations. Finally, there is the separate but related issue: mass destruction weapons are currently identified as primarily indiscriminate weapons, thus their prohibition. If technology succeeds in making them discriminate weapons—causing target destruction but not mass destruction—it is quite probable that, due to combat practice, states will identify them as primarily conventional.

Limitations of Arms Limitations

Arms control measures, accordingly, are not guarantees that force, or weapons, or hostilities will not occur or be used. In fact, the United States has taken the position with respect to these agreements that unless there is an express provision that the agreements apply during armed conflict, they will terminate or be suspended during conflict. If this position is taken, then the arms control agreements operate exclusively in peacetime. Certainly it is difficult enough to conduct compliance and verification checks during peacetime when these procedures are largely limited to "national technical means," e.g. electronic and satellite observation. But during wartime, compliance as a policy becomes meaningless. It is, practically speaking, nearly impossible.

If war breaks out, the applicable law becomes the law of war. As mentioned earlier, this law is to be found in treaties and agreements and in the customary practice of states. It tends to be formulated in very general terms, but its application and development depend upon reciprocal treatment by the belligerents during wartime.

Custom and usage tend to flow into the expectations and patterns of behavior of states, and when states exhibit, "subjectively," their intention to be "legally bound" by such law it becomes a part of the customary international law embraced in the law of war. Customary international law is therefore said to encompass both this subjective element and a separate objective element—exhibited in state action and behavior—revealing an expectation that there are restraints and that they will be enforced or applied. Enforcement of such law is largely through the use of counterforce; and to a lesser extent, during armed conflict, through allegations of war crimes, claims ultimately to be made for compensation and reparations. Perhaps in the future enforcement may be in large measure through international public opinion and through international forums if operative. Customary international law is thus to be distinguished from conventional or treaty law. Treaty law is enforced and applied as the law of the agreement itself. Such law shapes the arms-control agreements, while much of the law of war—the restraints on the use of weapons or force during hostilities or warfare—is based upon customary international law. Moreover, treaty law is established by agreement, most often through the bargaining processes among states, taking into account their security and defense interests.

These articles may be supplemented by others in the United Nations Charter, but Article II (4), in particular, establishes the fundamental norm against the use of force. However, its language, while broad in nature, does not deny states the right to use force to protect their interests.

Recent proposals for refraining from the use of force or even from the first use of force have been at best redundant. Such proposals, which are made from time to time by the Soviet Union, provide no additional law for outer space or elsewhere concerning restraints to be imposed upon aggression, the initiation of force for hostile purposes, and so on. The United Nations Charter requires no further elaboration in these areas and may be weakened by a presumption that new law was needed.

Further interpretation has been given to the charter in the Declaration of Principles of International Law Concerning Friendly Relations and Cooperation Among States in Accordance with the Charter of the United Nations, or the declaration, for short. Though the declaration has proved to be somewhat ambiguous, at best incorporating the accepted customary international law, it is useful as written law that may be invoked readily without dispute over language.

Just as the arms-control agreements provide no guarantee of protection against the use of force, or even against the use of the weapons that are covered in the agreements, a similar difficulty arises with the declaration, which we might assume is to protect us from aggression. And that difficulty is the inability of states to settle on what constitutes aggression. The resolution of the U.N. General Assembly that defines "aggression" fails to overcome this. Such lack of accord is inevitable when states justify their use of force, often by the general proposition of self-defense. This led to the enigma that baffled the philosophers Thomas Aquinas, Grotius, and others: the "just war." Because the use of force is condoned in defense of allies or of states requesting military assistance, the Soviet Union claimed a right to intervene with its forces in Hungary and Czechoslovakia, insisting that it had been invited in by the legitimate government. In fact, the Soviet Union has drawn up nonaggression and mutual defense treaties that make such consent to the use of force far easier to identify because it is given in advance by the signatories.

The Conduct of War

The last of the major areas to which law concerning the use of force is applied is that of military activities during warfare. The law of war is largely the outcome of the customs and practices of states during armed conflict. It has developed as a restraining influence, promoted by pragmatic considerations, i.e. the excessive use of military force does not achieve objectives, may lead to increased resistance by the enemy, may disturb the political objectives for which the war was fought, and unquestionably results in waste of scarce resources and thus in future weaknesses by the side engaging in force over zealously.

The fundamental principle from which the law of war has been derived is military necessity. This legal principle balances the demand of the military commanders and civilian policymakers to pursue the necessities of war, against the moderating demands made by citizens and the global community that humanitarian treatment be afforded those not directly involved in war.[2] Humanitarian treatment is, accordingly, identified as a part of the law of war. It is currently embodied in and applied through the Geneva Conventions of 1949 and through the customary international law supplementing those conventions that protects victims of armed conflicts, prisoners of war, armed forces personnel who are *hors de combat* (unable to continue fighting), civilians, medical and clerical personnel, and similarly uninvolved persons. The destructive aspects of war are also moderated through the law of war applicable to the conduct of hostilities. This law, in accordance with the principle of military necessity, designates that only legitimate military targets may be attacked. Further restrictions limit the degree of force to that which is proportionate to the military objectives to be gained and prohibit excessive force, i.e. that which would cause unnecessary suffering or result from indiscriminate use.

A detailed examination of the law of war and its development is beyond the scope of this paper, but, generally speaking, the law is shaped by the reality of warfare and the practice of states. No belligerent wants to be subjected to indiscriminate or wasteful destruction of its own resources and citizenry, or to be compelled to waste its own armaments, hence the law develops through reciprocal tolerances. Clausewitz's dictum, now an adage, defines war as an instrument of politics. The dictum has been superseded in a sense that other hostile measures have been adopted because their legitimacy cannot be freely attacked under international law and the United Nations

Charter as war. But Clausewitz's perspective embodies the economic, i.e. to avoid wasteful use of resources and manpower because to do so would diminish the rewards that are sought and impair the attainment of other objectives. The value of Clausewitz's insight is twofold: it affords a supporting counterpart to the legal principle of military necessity, and it is applicable to the actions of states during both war and peace.

Outer Space Demilitarization

The emerging law with respect to outer space is conditioned by the law of war as outlined above. As noted, the Soviet Union has made proposals for the nonuse of force or no first use of force that provide very little in the way of controls, primarily because they cannot be verified for compliance. They could be breached or abandoned in accordance with the interests of the nations involved.

The Soviet Union has recently proposed that outer space become a weapon-free zone that, according to its spokesman, would lead to the demilitarization of outer space. However, the proposal is deficient in that it does not prohibit the use of force or weapons in outer space, or eliminate the right of self-defense when a nation faces an attack. It does not cover weapons that might, at any time, be launched into outer space, and does not cover weapons modernization or research. For this reason, the Soviet proposal does not really prohibit antisatellite weapons because, to do so, would amount to denying states their right of self-defense. Its effectiveness is totally dependent upon all states ratifying the agreement without substantial modification. Also, the proposal, even if it were accepted by all states, does not provide for compliance verification, and even if it did, the present facilities of states are inadequate to assure such compliance.

To demilitarize outer space or create a weapons-free zone in outer space requires more than treaties, declaration, promises, or arms-control agreements that can disintegrate during wartime. None of these attempts confront the present problems generated by a competitive struggle for influence, power, and prestige among nations.

Law and Order in Space

The creation of effective controls for clearing space of weapons requires a global community whose members are dedicated to maintaining their own security and order. It requires the fulfillment of the fundamental goals of the United Nations Charter, but this has remained as remote as it was when the charter was adopted. To expect that through arms-control agreements or other fragmentary attempts to regulate the use of force states will be effectively restrained from that use is illusory. They may be deterred to some extent, perhaps even substantially. But the meaning of deterrence must include the use of any force that could begin a chain reaction that would lead to a general nuclear war.

The practice of states in outer space reveals their expectations that they must and will maintain their security. This includes not only their activities in outer space but also the security consequences of those activities on Earth. Because the global community provides no effective means for ensuring the security of its members, both the United States and the Soviet Union (as well as other nations) have continued to test, upgrade, and produce their weapons. Because each nation determines for itself how its weapons may best be improved, there may be unintended consequences for the entire system—consequences no state may have desired.

The right of the contracting states to pursue modernization, i.e. improvement, including research, development, and, depending upon the treaty, testing, in effect shifts the arms race from simply a quantitative race to a technological one. Old or technologically less-advanced weapons must be replaced. Moreover, this right conforms to the perspective of the two major rivals: neither can monitor compliance with such activities as "modernization." Accordingly, it would be incompatible with their security perspectives to impose prohibitions on themselves when they cannot be guaranteed that the other side will not cheat.

The security balance among the great nations has been in the past, and no doubt will continue to be, safeguarded by a rationally conceived arrangement of spheres of influence and mutual deterrence. The advent of nuclear weapons has only strengthened the stability of this balance. But a weakness in the balancing process—especially in regard to outer space—lies in conventional weapons that cannot be covered by verifiable agreements. It also lies in the difficulty if not impossibility of controlling and preventing the impermissible use of force when there are weaknesses in effective communications.

This brings me full circle. The diplomatic exercise that led to a "definition" of aggression in the United Nations General Assembly, which was discussed earlier, is instructive. While all sides sought to sharply limit impermissible uses of force, the definition they adopted would, under United Nations Charter practice, leave open for each to determine, largely for itself, when to use force and what standards to apply. Even if force for some purposes is declared impermissible, it can be effectively restrained only by those nations that have the power—through force or coercion—to impose their will. This of course returns us to the dominance of the powerful over the weak and leaves at deadlock the actions of nations with equal power.

Past experience, I believe, indicates that the Soviet Union is not influenced by unilateral measures undertaken by its opponents to promote disarmament.

Its perceptions of what is needed for its security differ from ours and—for whatever reason—have led to an awesomely powerful military force. The security environment, arising from the interdependence of U.S. and Soviet strategies, is not conducive to effective and enforceable law, nor to compliance with the agreements that have already been made. The practice of the United States and that of the Soviet Union involve confrontational positions: adversarial roles or rivalry on nearly every major event occurring throughout the world. Each perceives the other as aggressor. Until fundamental practices and attitudes change, it is unlikely that arms-control agreements for outer space will provide us with the mutual security we seek.

Notes

1. Chapter VI, "Pacific Settlement of Disputes," Article 33 of the United Nations Charter.
2. The wording used to "define" or characterize these principles in the Department of the Navy's *Law of Naval Warfare*, 1959 edition, is: (a) *Military Necessity*. The principle of military necessity permits a belligerent to apply only that degree and kind of regulated force, not otherwise prohibited by the laws of war, required for the partial or complete submission of the enemy with the least possible expenditure of time, life, and physical resources. (b) *Humanity*. The principle of humanity prohibits the employment of any kind or degree of force not necessary for the purpose of war, i.e., for the practical or complete submission of the enemy with the least possible expenditure of time, life and physical resources.

Part II

Civilian Activities in Space

6

The Space Station: Mankind's Permanent Presence in Space

Hans Mark

On 25 January 1984, in his annual State of the Union message, President Reagan announced a new national commitment to take the next important step in space operations.[1] The president directed the National Aeronautics and Space Administration (NASA) to plan a permanently manned orbiting space station that would be deployed within the decade. He also invited our friends and allies around the world to join with the United States in this new venture. His initiative marks the beginning of a new era in space, but it is also the end of a lengthy process of inspiration, technical planning, and political persuasion. It is the story of many people making their individual contributions one step at a time, which led finally to the president's commitment.

Early Concepts

The space station is not a new idea. In 1869 a Boston clergyman, Edward Everett Hale, wrote a story that appeared in the *Atlantic Monthly* called "The Brick Moon."[2] The story is a whimsical fantasy about the creation of a large Earth-orbiting sphere, built of brick, 200 feet in diameter, in which people lived and worked. The first quantitative work, though, was done by H. Oberth[3] and K. Tsiolkovsky.[4] They in turn strongly influenced two young engineers, Robert Goddard in the United States and Wernher von Braun in Germany. It was the work of Goddard and von Braun and their followers that made rocket propulsion feasible and thereby space station construction possible.

I first read about space stations more than 40 years ago in a book by P. E. Cleator called *Rockets through Space—The Dawn of Interplanetary Travel.*[5]

Cleator's book is a good technical review of the status of rocketry in the mid-1930s, but it also has important sections containing informed speculations about what can be done with the rocket technology he describes. The orbiting space station is a recurrent theme in the book. Cleator was particularly interested in the possibility of constructing a space station, or an "Outward Station" as he calls it, to be used as a staging base for trips to the planets. The concept of a staging base is today one of the principal reasons why the commitment to construct a space station has been made.

Rocket Developments

World War II saw great strides in the advancement of rocket technology. The technical tour-de-force of the Allies was the creation of nuclear weapons while the perfection of large, long-range rockets was the important technical achievement of the Germans. Even before the war was concluded, there were some who saw that the combination of nuclear weapons and large rockets would lead to a completely new "intercontinental artillery," as John von Neumann called it, that would have profound effects on world politics. It was, of course, the push to create the rockets to throw nuclear warheads from one continent to another that eventually led to the first space launch vehicles. After the end of the war, both in the Soviet Union and in the United States, German technology was adapted for making intercontinental rockets, and both sides used captured German scientists and engineers to do the job.

In 1952, Wernher von Braun, who by then was working in the United States, wrote an article on space stations for *Collier's* magazine.[6] In this piece, he restated the old ideas of Oberth and Tsiolkovsky and asserted that the new rockets then being conceived would eventually make space travel possible. He also, incidentally, for the first time suggested the idea of a reusable spaceship, which he called a "space taxi," but which we now know as the Space Shuttle.

Unfortunately, the U.S. political leadership at that time did not appreciate the impact that the first artificial satellite would have on the public, thus von Braun and his colleagues were prevented from achieving this historic "first." It was left to the Soviets to take the epoch-making step into space by launching Sputnik I into Earth orbit in 1957. This event changed things, and von Braun and his colleagues succeeded in launching their first satellite on 31 January 1958, a few months after the Soviets' historic achievement. Once this event was behind us, the space race was on, and the late 1950s and early 1960s saw a series of space firsts in which the Soviets and the Americans vied for supremacy.

The Trip to the Moon

Shortly after his election to the presidency in 1960, John F. Kennedy recognized the importance of space operations as a symbol of national competence in technology. He also understood that it was not enough to beat the Soviets in small ways but, rather, that some kind of a quantum leap would be necessary. Accordingly, he approved a plan that had been submitted to him by NASA Administrator James Webb to "land a man on the moon and bring him back safely before the end of the decade." By any standard, this was a tall order, but on 25 May 1961, President Kennedy gave the go-ahead and the Apollo program was born.[7] It was recognized from the very beginning that a staging base of some kind would be necessary if the objectives of the Apollo program were to be realized. The early plans, which were partially formulated by Wernher von Braun, called for a staging from Earth orbit. Two NASA committees, one chaired by Dr. Bruce T. Lundin and another by Dr. George M. Low, recommended this method. A small orbiting station would be placed into Earth orbit, from which the vehicle destined to go to the moon would be launched. In addition to being a feasible way to go to the moon, this method of "Earth orbit rendezvous" had the advantage of automatically leading to the construction of an Earth-orbiting space station. Many proponents of this idea felt that such a space station would ultimately be the most important result of the trip to the moon.[8]

But things turned out differently. The "Earth orbit rendezvous" method was more expensive and would take longer to achieve than the alternative of the "lunar orbit rendezvous" proposed by Dr. John Houbolt of the NASA Langley Research Center.[9] Houbolt proposed that the lunar landing ship go to the moon from a small station placed in orbit around the moon and then return to the lunar orbiting station for the trip back to Earth. This method is, indeed, more efficient and requires less total energy than the trip from Earth orbit. Because of the pressures imposed by a tight time schedule and limited funding, Houbolt's method was adopted by NASA in the summer of 1962. The lunar orbit rendezvous was brilliantly effective for going to the moon, but it had one major drawback, i.e. there was no automatic legacy of an Earth-orbiting space station at the completion of the Apollo program. It is true that an attempt to remedy this situation was made in the early 1970s with the Skylab program: when the decision was reached to terminate the Apollo program upon the completion of Apollo 17, permission was obtained to modify the leftover hardware from the Apollo program to place a Skylab in Earth orbit.[10] Skylab, as it turned out, was actually something of an afterthought even though it was very successful in its own right.

The Post-Apollo Program

Even during the height of the Apollo effort, studies were conducted by NASA on the subject of space stations. These were not large undertakings, but they served the important purpose of keeping people interested in the problem and also permitted incorporation of the latest technology derived from other programs into the plans for the space station.[11] Late in 1968, NASA acting administrator Dr. Thomas O. Paine established a committee to look at the future of NASA and to develop what he called a "Post-Apollo" program. He anticipated a successful landing on the moon and felt, correctly, that he should have something ready to propose once this all-important milestone was passed.

Among the committee members were all the directors of the NASA Centers. Since, as it happened, I became director of NASA's Ames Research Center in Mountain View, California, in February, 1969, I joined the committee just as the serious work was starting. The chairman, Dr. George Mueller, then NASA's associate administrator for Manned Space Flight, was an extremely imaginative and forceful individual. Yet without doubt, the dominant personality on the committee was the same Wernher von Braun who had done so much to draw public attention to space stations sixteen years earlier with his *Collier's* article and who had led the development of the large liquid-fueled rockets during World War II. (Von Braun was, at the time, director of NASA's Marshall Space Flight Center in Huntsville, Alabama.) He had a very clear idea about what the Post-Apollo program should look like. He told us in strong and eloquent terms that the time had come to construct an Earth-orbiting space station and that the Post-Apollo program should be worked out around this central idea.

In addition to the space station, von Braun also thought it would be important to create a new transportation system to carry things and people back and forth to the space station. This transportation system would consist of completely reusable rocket ships (or rocket airplanes) that would, as Wernher put it, "shuttle" back and forth between the ground and the space station, as he had proposed in his 1952 *Collier's* article. He foresaw that a space station would generate enormous traffic in space and that the reusable space shuttle would eventually be the most economical way to handle the traffic. Von Braun succeeded in persuading us that a reusable rocketship was important. Thus, the Space Shuttle was born.

Why did von Braun want a space station? What were the arguments he made that ultimately persuaded the rest of us on the committee that we should go ahead and propose a space station program? There were, and still are, today, four major arguments in favor of constructing a space station:

1. A space station is a laboratory in Earth orbit on which a great many scientific experiments can be performed that would be impossible to do on the ground. Included among these are biological experiments to determine the reaction of living things to zero gravity; certain chemical, physical, and fluid processes that can be carried out only in zero gravity; and many astronomical and astrophysical observations for which the absence of an atmosphere is an advantage.

2. A space station is a repair and maintenance base from which important satellites could be reached, retrieved, and maintained or repaired. This was not as strong an argument in 1969 as it is today. The United States now has deployed in Earth orbit well over 150 satellites, many of which perform missions of great importance to our national security and others that are economically very important. Right now, all of these satellites are eventually discarded when they run out of fuel or when various components reach the end of their lifetimes. The space station and the associated orbital transfer vehicles will eventually make it possible to maintain and service all of these satellites while they are in orbit. Satellites will, therefore, eventually become "permanent orbital facilities" that will be repaired, maintained, and upgraded as required. The shuttle flight conducted in April 1984, during which a failed satellite—the so-called Solar Max—was retrieved, repaired, and redeployed foreshadows the kind of operation that will become routine once the space station is completely operational.

3. A space station is a necessary staging base for more ambitious missions that will be carried out in the future. I have already mentioned the importance of staging bases in connection with the Apollo program. In the coming years, we will return to the moon. My guess is that we will establish a small permanent colony on the moon before the year 2000. We will also want to conduct more ambitious missions, such as sending people to explore Mars and possibly to visit some asteroids. Such missions can be executed much more easily if the starting point—the base camp, if you will—is a space station that permits refueling and resupply before the journey to Mars or some other distant object actually gets under way.

4. A space station is a symbol of our national competence in high technology and in the exploration of the unknown. This point is important politically, and President Reagan underscored it by inviting our friends and allies around the world to join with us in the development and construction of the space station. Imaginative initiatives of this kind have often had enormous political impact that have value much beyond the funds expended on them. This argument is, of course, similar to the one used by President Kennedy to justify the moon trip in 1961.

These were the reasons we considered back in 1969, and they are still the reasons which justify the space station program today.

Shortly after the successful lunar landings in 1969, the proposal to build the space station and the associated space shuttle was formally made by

NASA. At that time, there were many other proposals developed by several other study groups: a Space Task Group working under the auspices of Vice President Agnew developed ambitious plans for trips to Mars and for extensive scientific exploration projects. There were other groups which contributed to these efforts, some working within NASA and others sponsored by the National Academy of Sciences. After extensive debate, the only major project that survived for serious consideration was Wernher von Braun's space shuttle, which was proposed by the committee headed by George Mueller. The space station was not forgotten but it was postponed. The essential reasoning went something like this: Although construction of the space station was the ultimate objective, the space shuttle was technically more difficult as well as expensive to build. Given very limited budgets and since the shuttle would be a critical linchpin for the construction and operation of the space station as well as other space missions, it was decided to defer construction of the station and build the space shuttle first. The shuttle would, therefore, tend to pace the whole space station program. It was also felt that once the shuttle became operational, it would be easier to convince people that a space station would be important. With the shuttle it would be possible to conduct operations that would make it clear what could be done with a space station and how it would greatly enhance future space operations. As things turned out, this is just how it happened.

The Space Shuttle and the 1970s

In 1969, the space shuttle was designed to be a totally reusable system. There was to be a large airplane, the booster, which was designed to take a small airplane, the orbiter, to an altitude from which it would take off and go on to Earth orbit. However sound this plan seemed, it was expensive. The totally reuseable shuttle would cost something like $12 billion in 1969 dollars. Money of this kind was just not available. Thus, the large booster was dispensed with and the familiar configuration of today's shuttle was substituted. In February, 1972, President Nixon approved the Space Shuttle program and NASA promised to do the following things:

1. to build a reusable spaceship;
2. to do this for $6.5 billion (1972 dollars);
3. to put 65,000 pounds into near Earth orbit;
4. and, finally, to fly for the first time in 1979.

In due course, these objectives were essentially achieved. We did build a reusable spaceship; it cost more than $6.5 billion—something like $9.0 billion in 1972 dollars; we have almost, but not quite, succeeded in meeting the 65,000-pound payload requirement; and we flew for the first time in 1981, not in 1979. The Space Shuttle program was an extremely difficult technical tour-de-force and it was, therefore, not surprising that the original optimistic objectives have not quite been achieved.

The Space Shuttle program was approved at about the same time that the Skylab program, which I have mentioned already, was started. The expectation at the time was that the shuttle would eventually be used to tend Skylab and so the dream of the permanent manned space station would be achieved some time before 1980. Unfortunately, things did not quite turn out as expected. The shuttle was late in flying and could not deliver new fuel supplies to Skylab. At the same time, sun spot activity higher than expected caused atmospheric resistance which in turn slowed Skylab, causing it to fall back to Earth in 1979. Thus, the first plan for a space station based on Apollo hardware was not realized.

During the decade of the 1970s, NASA's technical efforts were focused on the shuttle. Although the plans for the space station were never far from our minds, not much actual work was done in those years on developing ideas for the program. There was one important exception. In 1974, Professor Gerard K. O'Neill of Princeton University visited me at the NASA-Ames Research Center and described some of the ideas he was developing for the establishment of large colonies in space. O'Neill began thinking about such things in the early 1970s when the *Limits to Growth* ideas promoted by organizations such as the Club of Rome were very fashionable.[12] Both O'Neill and I were disturbed by the great popularity of this world view and especially by the fact that some influential political leaders were actually proclaiming that the Era of Limits was at hand. We felt that such thinking was ultimately very dangerous because the necessity to apportion limited resources would ultimately lead to authoritarian political systems. The only way to meet this threat, in our view, was to make it credible that the resources available to mankind were ultimately not limited. We felt that not only prosperity but human freedom itself depended on the continual expansion of available resources and of the constant broadening of human horizons. It was for this reason that O'Neill's work was important and why, in more recent years, it has had great impact. In the summer of 1975, O'Neill headed an intensive study effort at Ames to more quantitatively develop the idea of establishing large human colonies in space.[13] While the implementation of O'Neill's proposals is not something that will happen in the next decade, his work did indeed inspire a new look at ways in which human horizons can be expanded. The extreme pessimism of the early 1970s was, in due course, replaced by the more optimistic view of the future that is current today.

Planning for the Space Station

Although the space station proposed by President Reagan is very modest compared to what O'Neill had in mind, it will give us the ability to expand our operations in space and will, eventually, lead to the larger scale activities that O'Neill proposes. This is particularly so for various manufacturing processes in space, and, ultimately, the use of extraterrestrial materials for these processes. Thus, while O'Neill's work did not have any direct bearing on the space station, it was extremely important in helping to create the political atmosphere in which it became possible for President Reagan to make his proposals.

The Space Shuttle orbiter Columbia flew for the first time on 12 April 1981. This epoch-making flight marked the end of the Space Shuttle's long period of technical development and began the transition to shuttle operations. Almost a month before the first flight, on 13 March 1981, I was notified that I would become deputy administrator of NASA in the newly elected Reagan administration. I was also told that I would be working for Mr. James M. Beggs, who was then executive vice president of General Dynamics Corp. in St. Louis, Missouri, and who would shortly be named as the new NASA administrator. Shortly after we were told of our appointments to our new positions, I went to St. Louis to visit with Jim Beggs and make some plans for what we would do when we assumed our posts. We agreed that the time had come to revive the plans for the space station that were made back in 1968 and 1969 and to attempt to persuade the new administration to adopt them. Fortunately, Jim was completely familiar with these plans since he had participated in formulating them during his prior service with NASA as an associate administrator. We agreed that we would mention the space station proposal as the next logical step for NASA during our confirmation hearings before the U.S. Senate and use that forum to test the political waters. As things turned out, there was no negative reaction to the statements we made at our hearings on 17 June 1981.[14] Indeed, we received considerable encouragement from several important quarters, so we decided to go ahead and lay the plans for persuading the president to commit the nation to the space station program.

Our first move was to ask former NASA administrator Dr. James C. Fletcher to organize a committee to advise NASA on how to proceed. The committee's charter was very broad, starting with the perennial questions as to the importance of people going into space versus doing the job with machines, to very detailed points about possible engineering designs, to broad questions dealing with the current political environment. Among others, the committee made one completely unequivocal recommendation: If a space station is to be constructed, it should be permanently occupied by people. In this way only could the full potential of space operations from the space station be realized. The committee also recognized that people in space do have political impact and that this factor should not be ignored.

The next step was to establish a Space Station Task Force charged with outlining the technical requirements and looking at some of the configurations that might satisfy them. This effort was led by Philip E. Culbertson and John O. Hodge. The work of this group made it possible to answer the myriad questions that were asked about the program. Several NASA centers—the most important being the Johnson Space Center and the Marshall Space Flight Center—also devoted strong efforts to planning for the space station. These efforts yielded the comprehensive plan for a permanently manned orbiting space station that ultimately became what the president proposed to Congress in January 1984.

Opposition to the Space Station

While it was extremely important to lay the technical foundations, this turned out to be relatively easy in view of the large body of work that already existed. The more difficult part of the effort was to persuade the political leadership that the space station should be built. There already was considerable opposition to the idea, especially from the scientific community and from the leadership of the military establishment. Their arguments went something like this: (1) The presence of people in space is unnecessary for the conduct of either scientific or military operations; these could be met most effectively by using automated spacecraft.[15] (2) The space station would divert from other, higher-priority space programs that both the scientists and the military believed to be more important. The tacit assumption made here was that, somehow, the total amount of money that the country spends on space activities is constant—that is, space operations are a zero sum game—and that money spent on one space project will not be spent on another.

It was in overcoming these arguments made by the opponents of the space station that Jim Beggs displayed his brilliance as a politically astute technical manager. He recognized early that there was really only one man who had to be persuaded: The president of the United States. In my fifteen-year experience in NASA I watched four NASA administrators deal with the problem of establishing proper relationships with the White House. Under normal circumstances, NASA is not a very important agency from a political viewpoint and, thus, NASA administrators normally do not have ready access to the president and to his closest advisors. Jim Beggs overcame the problem of access by using the popular attention that the spectacular flights of the Space Shuttle had attracted. The first landing of Columbia at Edwards Air Force Base in California on 14 April 1981 was witnessed by several hundred thousand people, and there was extensive worldwide press coverage of successive shuttle flights. Somehow, he felt that this great public interest must be used to draw the president's attention to the future of the American space program.

The Space Station Debate

The first step was to persuade the president to witness a shuttle landing himself so that he could see what an impressive technical achievement the shuttle was and what popular enthusiasm it generated. An opportunity to do this presented itself in due course, and President Reagan attended the landing of Columbia as she returned from her fourth mission on 4 July 1982. As Jim Beggs had anticipated, it was a festive occasion combining our national holiday with the excitement of the landing itself, and there is no doubt that the event captured President Reagan's interest. A crowd estimated at more than 500,000 people attended the event. The president made a short speech after the landing in which he promised "to use the capability of the shuttle to establish a more permanent presence in space." While this statement fell short of a commitment to build the space station, it was certainly a step in the right direction and left no doubt that the first move in engaging the president's attention was an unequivocal success.

Although some people were disappointed that the president did not make a commitment to build the space station in the summer of 1982, NASA was not, at the time, pushing very hard to have a formal announcement made. Many of us felt that we were not quite ready to make a firm proposal based on sound technical plans. Also, we felt that it would be important to have some time to see what could be done to persuade those of the president's senior advisors opposed to the space station proposal to modify their views. Accordingly, the space station effort in the last six months of 1982 and for the first half of 1983 consisted of two separate but related activities.

The Space Station Task Force performed several intensive studies of the requirements the space station would have to fulfill. Having done that, the Task Force then sponsored several conceptual design studies that developed the ideas for the basic architecture of the space station.[16] The space station would consist of a manned station made up of several different modules designed for various purposes, such as habitation, power production, stationkeeping, and experiments. In addition to the central manned unit, there would also be two unmanned platforms that would, so to speak, fly in "formation" with the space station itself. These platforms would be used when the proximity of the manned habitation modules would be harmful. This would include, for example, experiments requiring very accurate positioning where people moving around in the habitation modules would create unnecessary interferences, or for experiments with hazardous materials that could endanger the manned space station unit. The two unmanned platforms would, of course, be tended by humans; people from the space station would visit them occasionally for repair, replenishment, and upgrading. The space station contemplated would house six to eight people on a permanent basis, with the average residence time for a crew member being about three months. If the current plan is executed, then the space station would be assembled in Earth orbit and activated in 1991 or 1992, consistent with the president's 1984 proposal that the space station be constructed within the next decade. The estimated development cost of the space station is about $8 billion in 1985 dollars.

While the technical work was proceeding in a satisfactory manner, our efforts to persuade the opponents of the space station within the administration did not succeed. The many discussions that took place with senior members of the president's staff were often very useful to us in sharpening the arguments in favor of proceeding with the space station. While there were changes of position among the opponents of the space station, none of these resulted in unequivocal support for NASA's space station proposal.[17] We recognized, therefore, that we once again faced the same situation with which previous NASA administrators had to cope in the initiation of the Apollo program (Webb in 1961) and the Space Shuttle (Fletcher in 1972). In both of these cases the president's senior advisors were opposed to NASA's proposals yet the administrator had somehow succeeded in engaging the president's attention directly.[18] All of this illustrated the wisdom of Jim Begg's original strategy of drawing the president's attention to the space program directly through the shuttle operations.

International and Commercial Interest

Another event occurred in the summer of 1983 that was helpful in creating the proper political climate for a positive decision on the space station. Late in 1982, it was proposed to send the Space Shuttle Enterprise to the Paris Air Show as an exhibit. The air show was scheduled for late May 1983, so the timing was excellent. The Paris Air Show is the primary aeronautical exhibition in the world, and many of us felt that the impact of the space shuttle on the European public and perhaps even on the European political leadership would be helpful in focusing the thinking of the administration on the future of the U.S. space program. There was some risk involved in the execution of this plan. While Enterprise was not a shuttle vehicle configured for space flight—she was built as a mock-up to iron out various production problems and to be used in the approach and landing tests to 1977—her absence in Paris would, nevertheless, be a severe blow to the program. Enterprise was scheduled for use in the "form, fit, and function" tests that had to precede activation of the new launch pad on the West Coast (at Vandenberg Air Force Base) and the second launch pad on the East Coast at Kennedy Space Center. Perhaps even more important, we had only one Boeing 747 configured to carry the Space Shuttle orbiter. Loss of the shuttle carrier aircraft would have a serious impact on the future flight schedule since the aircraft was necessary to move the orbiter from the various landing sites to the launch site at the Kennedy Space Center. Eventually, it was decided to accept these risks and to send Enterprise to Europe.

There is no doubt that the European visit of the Enterprise was an unqualified success. Whenever the Enterprise and the shuttle carrier aircraft landed (Germany, France, Italy, and England), hundreds of thousands of people came to see the vehicles. Furthermore, during the extensive overflights of various regions in Europe, many more thousands watched the Enterprise and her carrier aircraft in flight. We estimated, conservatively, that about 2.5 million people in Europe saw the airplane. Perhaps even more important, the political leadership in Western Europe became aware, probably for the first time, of the enormous popular appeal of space flight. This appeal was further confirmed later in 1983 when the first European astronaut, Dr. Ulf Merbold, a German physicist, flew on the initial Spacelab mission in November, 1983. Dr. Merbold quickly became a folk hero in Western Europe and received very favorable coverage in the European press. It was also arranged during the Spacelab mission to have President Reagan and Chancellor Helmut Kohl of the Federal Republic of Germany talk to Merbold and to each other through a direct telephone voice link-up on Columbia. This event also served to draw the attention of the highest levels of the political leadership on both sides of the Atlantic to the potential of space flight. All of this helped set the stage for a positive decision on the space station by President Reagan during the final months of 1983.

In addition to the undoubted political impact of recent space operations, there was another factor that had important influence on the thinking of the president and some of his principal advisors. This was the interest of influential people in the private sector about the commercial potential of space operations. As early as 1947, Willy Ley, in his book *Rockets and Space Travel*,[19] expressed the strong hope that a space station (or the "Terminal in Space," as he calls it) would have very important commercial value. In the 1960s, the commercial viability of satellite communications was clearly demonstrated and today this industry is one of the fastest growing segments of the private sector. So far, communications is the only profitable space operation, but there are many who believe that other profit-making space enterprises are just over the horizon.

Perhaps the best example is the zero-gravity continuous-flow electrophoresis experiment that has been carried out on several Shuttle flights.[20] This experiment is a private investment by McDonnell Douglas Corp. and a subsidiary of the Johnson & Johnson pharmaceutical house carried out under a Joint Endeavor Agreement (JEA) with NASA. What has been learned so far is that the continuous-flow electrophoresis process works much more effectively (by more than a factor of 1,000) in zero gravity than it does on the ground. This means that certain very valuable materials can be made in commercially viable quantities in space, whereas they probably cannot be made on the ground. It is too early yet to tell whether this particular operation will ultimately be of commercial value, but there is no doubt that some people are already willing to put money on the line because they believe that it will be important. On 3 August 1983, President Reagan met with a group of chief executives of major corporations who told him that there was, indeed, commercial potential in space and that government investments in space operations would pay off in the future just as they have in the past. There is no question that this meeting had an effect on the president's thinking as he considered the space station problem.

The President Decides

In the summer of 1983, NASA's budget proposal for fiscal year 1985 was completed. It was decided to make a formal proposal to go ahead with a commitment to build a permanently manned space station, initially with $235 million. This figure was reduced later on, after intensive negotiations with the Office of Management and Budget, to $150 million, but that was still enough to get a strong start on the program.

During this same period, several meetings of cabinet officers and senior advisors of the president were held to discuss the space station. The president himself was present at some of these meetings. He listened attentively to the various viewpoints presented by those who opposed NASA's proposals and participated actively in the discussions. At the conclusion of this series of meetings, it was clear that the president would support NASA's proposals, and in mid-December, NASA was asked to supply a paragraph for the president's State of the Union Message, scheduled for 25 January 1984, announcing the commitment of the administration to the construction of the space station. It was significant that, as the announcement was formulated, an invitation was included to our friends and allies around the world to work with us on the space station program. The international commitment was largely the result of the intensive work that various representatives of NASA had done during the previous year to develop interest among the international community in participating with the United States on this important new effort.

With the president's commitment to initiate the space station program, the first phase of the effort was complete. There are still many hurdles to be crossed and many setbacks to be endured before the space station becomes a reality. There are technical issues and problems related to international participation in the space station program that must be resolved. Nevertheless, there is reason to believe that these obstacles will be successfully negotiated and that the space station will be built as the president has proposed. A new era has begun, and it is fitting and proper that the United States of America is leading the effort.

Notes

Dr. Mark's article is based on his 1984 Bauer Lecture to the Aerospace Medical Association. It has appeared in the Association's journal, *Aviation, Space, and Environmental Medicine*, and is used here with the kind permission of Dr. Mark and the Association.

1. President Ronald Reagan, State of the Union Message to the Congress, 25 January 1984.
2. Edward Everett Hale, "The Brick Moon," *Atlantic Monthly* 24 (October-December 1869): 451–688. This story was published later as a book, *The Brick Moon and Other Stories* (Boston: Little, Brown, 1899).
3. Hermann Oberth, *Die Rakete zu den Planetenraumen* (Munich: R. Oldenbourg, 1923); Konstantin E. Tsiolkovsky, "Issledovanie mirovykh prostranstv reaktivnymi priborami" (A study of Atmospheric Space Using Reactive Devices). This 1903 paper set forth his theories on space travel, a subject upon which he had been working since 1896.
4. Robert H. Goddard, "A Method of Reaching Extreme Altitudes," *Smithson. Misc. Coll.* LXXI, No. 2, 69, 1919; Wernher von Braun's early contributions are best described in a book by Walter Dornberger, *V-2* (New York: Bantam and Viking, 1954).
5. P. E. Cleator, *Rockets through Space: The Dawn of Interplanetary Travel* (New York: Simon & Schuster, 1936).
6. Wernher von Braun, "Crossing the Last Frontier," *Collier's Magazine*, 22 March 1952. This article was one of a series published by various space experts in *Collier's*. These articles were expanded and republished in a collection by Cornelius Ryan, *Conquest of the Moon* (New York: Viking, 1952).
7. President John F. Kennedy. Address before a Joint Session of the Congress, 25 May 1961.
8. Ivan D. Ertel and Mary Louise Morse, *The Apollo Spacecraft—A Chronology.* NASA SP-4009, 1969.
9. John Houbolt, "Lunar Orbit Rendezvous and Manned Lunar Landing," *Astronautics* 7 (April 1962):20–29.
10. W. David Compton and Charles D. Benson, *Living and Working in Space: A History of Skylab.* NASA SP-4208, 1984.
11. "The Needs and Requirements for a Manned Space Station." NASA Report prepared by the Space Station Requirements Steering Committee, Charles J. Donlan, Chairman, 15 November 1966.
12. Donella H. Meadows, Dennis L. Meadows, Jorgen Randers, and William W. Behrens III, *The Limits to Growth* (New York: Potomac Associates Books, 1972).
13. Gerard K. O'Neill, *The High Frontier* (New York: William Morrow, 1977).

14. U.S. Senate, Committee on Commerce, Science and Transportation, *Hearings: Nominations—NASA* (Serial No. 97-48, 17 June 1981). See especially page 22 of the hearings.
15. R. Jeffrey Smith, "Squabbling over Space Policy," *Science* 217 (23 July 1982):331.
16. Space station program description document, NASA-Internal Papers, 1983.
17. M. Mitchell Waldrop, "The Selling of the Space Station," *Science*, 24 February 1984, p. 793.
18. Hugh Sidey, "The Presidency," *Time*, 14 November 1983, p. 69.
19. Willy Ley, *Rockets and Space Travel* (New York: Viking, 1947).
20. Jerry Grey, "Investing in Space: Now, Soon, or Later?" *Aerospace America* 22 (April 1984):90.

7

U.S. International Space Activities

Marcia S. Smith

Introduction

During the last twenty-five years the United States has concluded over a thousand agreements with more than a hundred countries for joint space programs. The activities are sponsored primarily by the National Aeronautics and Space Administration (NASA), but other government agencies, including the Agency for International Development and the Department of Commerce, have also been involved.

Although the vast majority of these international ventures have been tremendously successful, difficulties have developed in a few cases which have at least temporarily cast doubts on the efficacy of U.S. international space policy. For example, the unilateral withdrawal of the United States from the joint U.S.–European Space Agency International Solar Polar Mission in 1981 led to official expressions of discontent to the State Department by several of the countries involved. In addition, there is increasing concern about the transfer of U.S. technology to other countries, which may hurt U.S. economic or national security interests.

In the broader world context, the United States participates in activities of the United Nations Committee on Peaceful Uses of Outer Space (COPUOS), which has drafted five space treaties, four of which are now in force, and serves as a forum for a discussion of international space issues. In August 1982, COPUOS sponsored the Second Conference on the Exploration and Peaceful Uses of Outer Space (UNISPACE'82), which was held to demonstrate the benefits of space activities to developing countries.

U.S. International Space Policy

The U.S. space program has been international in character since its inception. The 1958 National Aeronautics and Space Act, which created NASA, states that space activities of the United States shall be conducted so as to contribute materially to cooperation with other nations in work done pursuant to the act, and in the peaceful application of the results of such activities. Directives by President Carter in 1978 and President Reagan in 1982 reiterated this policy, although the Reagan directive added that international space activities shall be pursued if they provide "scientific, political, economic, and national security benefits."

The Reagan language demonstrates the many facets of international space cooperation. Initially, the international component of NASA's program was pursued for two reasons: first, because of a genuine desire to involve other countries in exploring the new frontier of space; and second, because NASA required worldwide tracking locations and wanted to generate a climate in which other countries would be predisposed to allow tracking sites to be established on their land. Later, as national budgets became more constrained, the economic benefits of cooperation also became important. Programs that were too expensive for the NASA budget could still be pursued if another country or group of countries was willing to provide some of the instruments, or even entire spacecraft. Similarly, by offering free launch and tracking and data acquisition services, U.S. scientists might be invited to fly instruments on a foreign satellite or at least receive data. More recently, under the Reagan administration, national security has become a watchword in virtually all technology areas, and space is no exception.

Types of Cooperation

There are many different types of cooperative programs, primarily sponsored by NASA, ranging from the exchange of scientists to joint development of spacecraft. A summary of NASA's cooperation programs through January 1983 is provided in Table 1.

One of the largest categories of cooperation involves the reception of data gathered by U.S. satellites, especially meteorological and remote sensing satellites. The NASA-developed Automatic Picture Transmission station for receiving data from polar-orbiting weather satellites is in common usage throughout the world, so much so that agreements are no longer signed for their use. The equipment is available through several international outlets.

The direct reception of land remote sensing data from Landsat satellites requires signed agreements. At the end of 1982, eleven foreign Landsat ground stations were in operation at the following locations: Argentina, Australia, Brazil, Canada (2), India, Italy (owned by the European Space Agency, ESA), Japan, South Africa, Sweden (owned by ESA), and Thailand. In addition, Indonesia and China have signed agreements to build ground stations, and several other countries have indicated interest. (Three agreements, with Chile,

TABLE 1
International Programs Summary

	Number Countries/ International Organizations	Number Projects/ Investigations/ Actions*
COOPERATIVE ARRANGEMENTS		
Cooperative Spacecraft Projects	8	33
Experiments on NASA Missions		
Experiments with Foreign Principal Investigators	14	73
US Experiments with Foreign Co-Investigators or Team Members	11	56
US Experiments on Foreign Spacecraft	3	14
Cooperative Sounding Rocket Projects	22	1,774
Joint Development Projects	5	9
Cooperative Ground-Based Projects		
Remote Sensing	53	163
Communication Satellite	51	19
Meteorological Satellite	44	11
Geodynamics	43	20
Space Plasma	38	10
Atmospheric Study	14	11
Support of Manned Space Flights	21	2
Solar System Exploration	8	10
Astronomy & Astrophysics	25	11
Cooperative Balloon and Airborne Projects		
Balloon Flights	9	14
Airborne Observations	12	17
International Solar Energy Projects	24	9
Cooperative Aeronautical Projects	5	40
Scientific & Technical Information Exchanges	70	3
REIMBURSABLE LAUNCHINGS		
Launchings of Non-US Spacecraft	15	95
Foreign Launchings of NASA Spacecraft	1	4
TRACKING & DATA ACQUISITION		
NASA Overseas Tracking Stations/ Facilities	20	48
NASA Funded SAO Optical & Laser Tracking Facilities	16	21
Reimbursable Tracking Arrangements		
Support Provided by NASA	5	48
Support Received by NASA	3	12
PERSONNEL EXCHANGE		
Resident Research Associateships	43	1,266
International Fellowships	21	358
Technical Training	5	972
Foreign Visitors	131	81,377

Source: *NASA Pocket Statistics*, January 1983, p. A-6.
*Completed or in progress as of 1 January 1983.
Summary for activities after 1983 is unavailable.

Iran, and Zaire, have lapsed.) Each country pays an annual access fee ($600,000 beginning in October 1982) and agrees to provide the United States with any data it needs. Originally, NASA was responsible for these agreements, but with the transfer of the operational Landsat program to the Department of Commerce's National Oceanic and Atmospheric Administration, the latter agency has taken over these activities.

NASA sponsored a program for assisting countries in interpreting remote sensing data from Landsat 1 and 2 and the Skylab EREP (Earth Resources Experiment Package) program. Approximately 50 countries participated, with NASA selecting individual experiments based on scientific merit. No exchange of funds was involved. Although the NASA program has been terminated, the Agency for International Development sponsors seminars throughout the world to demonstrate the value of remote sensing data and has established regional remote sensing centers in Thailand, Kenya, and Upper Volta.

A third category of cooperation concerns the launch of satellites. There are two types of agreements—cooperative and reimbursable. In a cooperative launch, there is no exchange of money but rather an exchange of data and services. For example, the United States might launch a satellite in which either the payload has been developed jointly or the foreign country has provided the entire spacecraft, but in return U.S. scientists are given access to the data. In a reimbursable launch, a country pays to have the United States launch a satellite (usually for communications). Italy has also launched several U.S. satellites, using U.S. launch vehicles, from its San Marco platform off the coast of Kenya.

The programs probably represent what commonly comes to mind when discussing "cooperative" space activities in which both sides contribute to the development of a space program and analysis of the data. Throughout the history of the space program, most of these missions have been U.S. spacecraft on which the United States has offered foreign scientists the opportunity to fly experiments, although in a few cases the other country has provided the entire spacecraft for launch by the United States (such as the Helios spacecraft built by Germany for solar studies). Also, the Soviet Union has launched a few spacecraft which carried U.S. experiments. With the advent of independent launch capabilities by other countries (notably the European Space Agency) the situation is now reversing to some extent, with opportunities for U.S. scientists to fly experiments on foreign satellites at a time when the U.S. space science program is constrained by the budget situation.

As is evident, U.S. international space activities are extensive and cannot be treated in detail here. A few examples of cooperative endeavors, both successful and not so successful, are provided below.

Case Examples of Successful Cooperative Programs

The U.S.–USSR Apollo-Soyuz Test Project

The most visible example of international cooperation is the 1975 Apollo-Soyuz Test Project (ASTP) with the Soviet Union. The United States launched a three-man crew in an Apollo spacecraft to dock with a two-man crew in a Soviet Soyuz. The five men exchanged pleasantries and gifts and conducted several scientific experiments in the two days that the ships were docked together. After undocking, the two ships conducted docking and eclipse experiments for two days. The Soyuz then returned to Earth while the Apollo remained in orbit for three more days to conduct additional scientific experiments.

The ASTP mission was notable primarily because it involved cooperation between two traditional space competitors. There had been little cooperation between the United States and Soviet Union through the 1960s, although agreements had been signed in 1962 and 1965. In 1969, NASA initiated talks for new cooperation, which led to a 1970 agreement to study the possibility of a joint manned mission. In 1972, an accord that led to ASTP was signed by President Nixon and Soviet Chairman of the Council of Ministers Kosygin.

The ASTP mission was criticized as an expensive method of demonstrating the warming of relations between the two countries and as a giveaway of advanced U.S. space technology. However, it is difficult today to point to a single example of new space technology being used by the Soviets that might have come from their experience with ASTP, with the exception of the remodeling of the Soviet mission control center to resemble the one at NASA's Johnson Space Center. While U.S. engineers did not take home invaluable technical expertise either, at least some new information was gained about the secretive Soviet space program. U.S. astronauts and support personnel were allowed to visit the Soviet launch facility at Tyuratam, and were able to study the Soyuz spacecraft (later noting that it was closer to a 1965-era U.S. Gemini spacecraft than to an Apollo). The West also obtained details of a 1971 Soviet space mission in which the three-man crew died during reentry—information that might not have become available without ASTP.

Another cooperative agreement signed in 1977 included the suggestion that a mission involving the U.S. Space Shuttle and the Soviet Salyut space station be pursued. In addition, the Soviets offered to fly U.S. experiments on several of their biological satellites. The first such mission had been successfully accomplished in 1975, with three more agreed upon in the 1977 agreement. U.S. scientists have reported excellent cooperation from their Soviet colleagues not only in the biological satellite programs, but also in the exchange of data from planetary flights (particularly Venus), and biomedical data from the long-duration Soviet manned spaceflights (the longest Soviet manned flight to date has been 211 days, compared to 84 days for the United States).

U.S.–Soviet cooperation cooled following the Soviet invasion of Afghanistan in 1979. The 1977 agreement expired in May 1982 and President Reagan decided not to renew it because of Soviet involvement in Poland. Two programs are being continued, however: the biological satellite mission mentioned above, and an international program for satellite-assisted search and rescue called COSPAS/SARSAT. A Soviet satellite carrying the first of a new generation of transponders capable of pinpointing aircraft and other vehicles in distress was launched in the summer of 1982, and within a short time was responsible for the safe return of a three-man airplane expedition that had crash landed in Canada.

Spacelab, the United States, and the European Space Agency

The European Space Agency (ESA) was established in 1975 as a merger of two organizations for developing launch vehicles (ELDO, the European Launch Development Organization) and satellites (ESRO, the European Scientific Research Organization). The United States had a history of cooperation with ESRO, which continued with the creation of the new agency. ESA now has eleven member countries (Belgium, Denmark, France, Germany, Italy, Ireland, the Netherlands, Spain, Sweden, Switzerland, and the United Kingdom), two associate members (Austria and Norway), and a technical agreement of cooperation with Canada.

Extensive cooperation has taken place between the United States and ESA, but only one case will be described here: ESA's development of a module called Spacelab for use with the U.S. Space Shuttle.

Under the terms of the original agreement, signed in 1973 with ESRO, the Europeans designed, developed, manufactured, and delivered one Spacelab module free to NASA, with the option for NASA to buy additional units. (The Europeans did this to stimulate their own aerospace industry and in the hope that NASA would buy additional units.) The pressurized unit fits inside the cargo bay of the Space Shuttle, providing a shirtsleeve environment for scientists to conduct experiments. The agreement benefited both sides by providing the United States with a needed component of its space transportation system, and European industry gained an opportunity to expand into new technological areas.

The first Spacelab was delivered to NASA in the spring of 1982. Although difficulties were encountered during the program, primarily caused by changing NASA requirements, which in turn led to redesign of certain Spacelab components, the overall program has been quite successful. The major ESA disappointment is that NASA has agreed to purchase only one additional

Spacelab unit, but the Europeans are promoting the concept of building permanent space stations by docking several Spacelab modules together. At the very least, the knowledge gained by ESA and by European companies will put them in a favorable position to participate in a manned space station program should the United States choose to pursue one, and possibly enable them to build space stations of their own.

The U.S.–Indian SITE Experiment

Another type of cooperation is exemplified by a 1975 program with India called SITE (Satellite Instructional Television Experiment). In this instance, the United States lent its ATS-6 satellite (Applications Technology Satellite 6) to India for one year so that country could perform experimental television broadcasts to remote villages. Programming was developed by the government of India to provide instruction in health, family planning, education, and agriculture in eight different languages (Hindi, Kashmiri, Bengali, Oriya, Marathi, Gujarti, Tamil, and English) directly to small receivers in 2,200 villages; another 2,000 to 3,000 villages were able to receive the broadcasts through secondary transmission methods.

The experiment was considered a tremendous success, and led to India's developing satellites to continue these activities. Launched by the United States on a reimbursable basis, India's satellites have continued the television broadcasts to remote villages, as well as provided point-to-point communications and meteorological information.

Where the Problems Are

The vast majority of cooperative programs have been extremely successful, but it would be remiss not to mention a few of the difficulties that have arisen in recent years. By and large, the problems have not been caused by the U.S. agencies involved in program implementation but, rather, by budgetary constraints within NASA, new concerns about technology giveaway, and U.S. policy regarding launch assurances for foreign countries.

The International Solar Polar Mission

In 1979, NASA and ESA signed a memorandum of understanding to conduct a program called the International Solar Polar Mission (ISPM, formerly called the Out-of-the-Ecliptic Mission) for launch in 1983. The plan called for each agency to provide one spacecraft for studying the polar regions of the sun. To achieve this goal, the spacecraft would have to fly out of the plane of the ecliptic (in which the planets orbit the sun), requiring a great amount of energy. Thus, the mission envisioned sending the two probes toward Jupiter where the giant planet's gravity field would be used to swing the spacecraft up and out of the ecliptic and back toward the sun. With two spacecrafts, simultaneous readings could be made of both poles. In addition to providing one of the spacecraft, NASA agreed to launch both of them, providing tracking and data acquisition services, and give ESA a radioisotope thermal generator (RTG) to power its spacecraft.

The program encountered funding problems from the very beginning. NASA requested ISPM as a new start in its FY 1980 (fiscal year 1980) budget, and although it was approved by Congress, the chairman of the subcommittee of the House Appropriations Committee with jurisdiction over NASA programs in a letter to the NASA administrator in December 1979 directed NASA to postpone the mission for two years because of: (1) problems with the development of the Space Shuttle, which was to be used as a launch platform for the spacecraft, and (2) committee concern that the upper stage which was to send the spacecraft on their journey (the inertial upper stage) would not be adequate for the job. The committee believed NASA should develop a high-energy upper stage (Centaur) instead. Although NASA was not convinced by these arguments, the agency subsequently postponed the launch of ISPM to 1985 as a budget move.

Next, in the spring of 1980, the House Appropriations Committee recommended in the FY'80 supplemental appropriation bill that ISPM be terminated and some of the money that had been appropriated for FY'80 be rescinded. In response, several member countries of ESA formally protested to the State Department, which then came to the aid of the program in Congress. The program was saved, but only temporarily.

The final blow to ISPM came in 1981 when President Reagan submitted his first budget request to Congress; it recommended termination of the U.S. spacecraft for ISPM, although support for the ESA spacecraft was maintained (launch, tracking and data-acquisition services, and the radioisotope thermal generator). The action was taken unilaterally, without prior consultation with ESA. At the same time, NASA decided that it would develop the Centaur upper stage instead of the IUS after all, which required changes in the ESA spacecraft.

Once again protests were lodged with the State Department and NASA, but this time to no avail. ESA Director General Erik Quistgaard, in testimony before the House Science and Technology Committee, reminded the Members that ESA had chosen ISPM over several other potential scientific programs because of the value it attached to cooperation with NASA, and that ESA has already spent $100 million on the program. Congress provided NASA with $3 million to keep the program alive, but the Office of Management and Budget (OMB) informed NASA that it would not support funding for the program in future years, although it would allow NASA to reprogram funds from other space science projects. After considering several options, including the possibility of purchasing a spacecraft from ESA, NASA finally decided that it could not afford ISPM without support from OMB.

ESA later decided that it would continue with its spacecraft plans, but has sought assurances from the United States that future budgetary actions would not terminate NASA's promise to launch the probe, furnish tracking and data acquisition services, and provide the RTG. Although ESA sought an earlier launch opportunity as recompense for U.S. cancellation of its spacecraft, the launch date for the mission has slipped an additional year, because the shuttle and Centaur upper stage will not be ready until 1986.

Assessing the long-term impact of the ISPM incident is difficult because no further ESA–NASA agreements have been signed since that time. ESA has continued to express interest in cooperating with NASA on space science programs, and possibly a space station, but it can be expected that the agency will want certain guarantees written into future memoranda of understanding to prevent unilateral cancellation of such programs, or at least to limit its financial liability in the event of cancellation. NASA faces the problem, however, of having to place contingency clauses in such agreements stating that continuation of the program is subject to appropriations, which make the programs vulnerable to annual reviews both by the administration and Congress.

Lack of Program Continuity

Another type of problem encountered in international programs because of budget constraints is continuity of effort. There are many examples of this, but only two will be cited here: Landsat activities and communications support to some Pacific islands.

The Landsat program provides multispectral data which can be used to survey natural resources, study the effects of pollution, assist in land use planning, and monitor crops and provide crop yield estimates. The United States has launched four Landsat spacecraft and one more is planned, but beyond that the future of the program is uncertain. The government wants the program to be taken over by the private sector, but because there is not yet a large enough market for land remote sensing data to guarantee a profit, there has been a dearth of proposals by the private sector to take over the program without government subsidies. Thus, users in foreign countries (as well as in the United States) are concerned that they will invest large amounts of money either in purchasing Landsat receiving stations or in training personnel to use the data, only to have the United States terminate the spacecraft segment.

Another aspect of the Landsat problem is that programs for providing direct U.S. support, formerly through NASA and currently through AID, for training foreign scientists in the interpretation and use of Landsat data are in danger of being terminated because of budgetary constraints.

The lack of continuity in U.S. programs is also evident in the communications area. As an example, the PEACESAT program has provided communications services to certain Pacific island nations since 1971 using the ATS-1 satellite. The satellite is now at the end of its lifetime, and there is no other satellite with which to carry on this service. Suggestions have been made for using one of the older INTELSAT satellites, but this would require installation of different ground stations. Another option is to use transponders on one of NASA's soon-to-be-launched Tracking and Data Relay Satellites, but this would require repointing one of the satellite's antennas, which NASA had indicated an unwillingness to do. Thus, this communications service may end in the near future.

Technology Transfer Issues

The debate over technology transfer has many components, including concerns about the amount of technology the United States is providing to other countries that then becomes an instrument of competition with U.S. companies; about technology that might be diverted to military purposes which is transferred to a country with which the United States has strained relations; and about technology that might be initially transferred to a friendly country but that ends up in the hands of a nation with which the United States is not so friendly, either through the capture of U.S. equipment or overthrow of a friendly government. An example of the latter is offered here.

In 1978, the Arab League, comprising 21 Arab countries and the Palestine Liberation Organization, decided to purchase a regional communication satellite system. Called ARABSAT, the satellite system consists of three geostationary communications satellites. The ARABSAT contract was won by Aerospatiale, a French company, teamed with Ford Aerospace in the United States. To provide Aerospatiale with satellite technology, Ford Aerospace required a technology export license, which must be approved by the Senate. In the fall of 1981, the State Department forwarded the request for such a license to the Senate. Unfortunately, the request came shortly after a stormy Senate debate over providing AWACS technology to Saudi Arabia, which had raised concerns about technology transfer to the Middle East.

When it was learned that the satellite system would be used by the Arab League, which includes some countries currently out of favor with the United States, several Members of the Senate felt that the export license should be denied. The State Department withdrew the request. Following several weeks of briefings by Ford Aerospace (which would have lost $20 million and 5,000 jobs), the Department of Defense, and the State Department, the senators relented after realizing that the technology was being transferred to France, not Arab countries, and that the satellites themselves would be in geosynchronous orbit, well out of the reach of any country. The request for an export license was reintroduced and not opposed. A major problem resulting from this controversy is whether U.S. companies will be considered reliable international partners in such endeavors in the future, or whether U.S. government approval will be required before a foreign company will sign such an agreement.

Launch Assurances

Yet another type of problem encountered in the international space business is that of launch assurances. In this case, it is a matter of what the United States will agree to launch for another country, even if that country is paying for the launch.

For example, the United States will not launch satellites for other countries that might adversely impact its national security; the assumption can be made that this would include military reconnaissance satellites, which may be desired by other countries. The United States also will not launch satellites that would violate its responsibilities under other international agreements; a case in point is the INTELSAT agreement.

The United States sponsored the formation of INTELSAT in the 1960s (and, in fact, INTELSAT is a good example of a highly successful U.S. international space initiative). One section of the INTELSAT charter states that signatories must agree not to launch satellite systems that might cause economic harm to the INTELSAT system—that is, INTELSAT and INTELSAT alone will provide international satellite communication services.

In 1967, however, Germany and France decided to develop a satellite system called Symphonie, for launch in 1974, capable of providing communications between Europe, Africa, the Middle East, and North and South America. When they asked the United States to launch the satellites, the response was that such a system would be competitive with INTELSAT, and the United States therefore could not provide launch services without violating the INTELSAT agreement. Eventually, the two Symphonie satellites were launched by the United States after Germany and France formally stated that the system was experimental, not operational. Although plans were already under way at that time for constructing a European launch vehicle capable of placing communications satellites in orbit, the Symphonie incident is often credited with providing the final impetus to the Europeans to construct their own launch vehicle so they would not have to depend on the United States. This resulted in the development of Ariane.

It should be noted that a number of regional communications satellite systems (such as ARABSAT) are now planned. To date, these regional systems have been approved by INTELSAT on a case-by-case basis to ensure that they provide regional rather than international service. A 1982 decision by INTELSAT to approve a European regional satellite system being built by the EUTELSAT consortium for only a five-year period, after which another review will take place, has raised the ire of several European countries. With the growing interest by individual countries and groups of nations in having their own satellite systems, INTELSAT will almost certainly have to review its position on competitive systems if it is to survive through the 1980s.

The United States may also have to review its policy regarding which types of satellites will be launched. While it cannot be expected to launch satellites that might negatively affect its national security, it is unclear as to which other categories of satellites might also be refused. This will become increasingly important as nongovernment entities such as Space Services, Inc., begin providing launch services.

The Politics of Cooperation

In addition to the practical, scientific, economic, and national security aspects of international space cooperation, there are definite political components, too. They can take several forms. As an example, NASA has been accused (with increasing frequency) of using international cooperation as a method for winning approval for its programs both within the administration and in Congress. During the late 1970s in particular, when NASA's space science budget started dwindling, an oft-repeated NASA theme in testimony before congressional committees was that other countries were participating in a particular program, and its cancellation would create international discord. The International Solar Polar Mission (ISPM) discussed earlier is a case in point.

This approach worked quite well until the Reagan administration, when such arguments seemed to lose their effectiveness with the executive branch (it was the administration, not Congress, that cancelled the ISPM). Nevertheless, NASA may have used the interest of other countries to help convince the administration to support construction of a space station as the next major NASA program. Approval has not been forthcoming, and NASA was reported to have been embarrassed by the fact that other countries (notably ESA and Japan) were publicly offering to cooperate with NASA in such an endeavor. Mutterings have been heard that NASA is attempting to stimulate international support to make it more difficult for the executive branch and Congress to reject the space station idea.

Another way in which politics influences cooperation is exemplified by the current state of U.S.–USSR cooperation. As noted previously, there are no cooperative agreements in place between the two countries because of a Reagan administration decision not to renew a series of scientific agreements in light of Soviet intervention in Poland. The only existing space programs are two that began earlier: the search and rescue satellite program and the biological satellite program. (The latter program is also interesting from a political point of view because of the pressure brought to bear on Congress by antivivisectionists who wanted the flight cancelled because of their concern about the treatment of the two rhesus monkeys involved in the experiment. Apparently they do not realize that the monkeys are being provided by the Soviet Union, so no U.S. action could prevent the flight.)

The International Setting

International politics can also play a role in space activities, although less so in specific programs (such as scientific probes) than in overall policy.

The United Nations Committee on the Peaceful Uses of Outer Space (COPUOS) serves as a forum for the discussion of international space issues. Through its Legal and Scientific Subcommittees and the Technical Subcommittee, recommendations are made for treaties and other types of agreements on international space issues. Five treaties have been developed by COPUOS, of which the first four are in force: the 1967 Treaty on Principles Governing the Activities of States in the Exploration and Use of Outer Space, Including the Moon and Other Celestial Bodies (the Outer Space Treaty); the 1968 Agreement on the Rescue of Astronauts, the Return of Astronauts, and the Return of Objects Launched into Outer Space; the 1973 Convention on International Liability for Damage Caused by Space Objects; the 1976 Convention on Registration of Objects Launched into Outer Space; and the 1979 Agreement Governing the Activities of States on the Moon and Other Celestial Bodies. The United States is a signatory to all but the last of these.

Other issues debated within COPUOS include principles governing direct broadcast satellites and remote sensing data, the use of nuclear power sources on spacecraft, the definition and delimitation of outer space, use of geostationary orbit, and technology transfer. In addition to the COPUOS activities, the Committee on Disarmament (CD), which is closely associated with the United Nations, is considering issues involving the "militarization" of space, and whether to establish an international satellite monitoring agency to allow the United Nations to monitor treaty compliance in times of crisis through use of reconnaissance satellite data. All of these issues were discussed in August 1982 at the Second U.N. Conference on the Exploration and Peaceful Uses of Outer Space (UNISPACE'82). The United States was one of 94 countries represented at the conference, which was sponsored by COPUOS to serve as a forum for demonstrating to developing countries the advantages of space activities (particularly in areas such as remote sensing and communications).

The two-week conference, rather than focusing on technical subjects, became embroiled in a controversy over whether the conference report should include language expressing the concern of the international community over the perceived "militarization" of space. The United States contended that the topic of militarization had been referred to the CD and therefore was inappropriate to discuss at UNISPACE'82, which was sponsored by COP-UOS. All the other countries thought it was quite appropriate because military uses of space can affect peaceful uses, and COPUOS has a direct interest in the topic since it originated the text of the 1967 Outer Space Treaty, which addresses the military uses of space in Article IV (banning the use of nuclear weapons or other weapons of mass destruction in space). After two stormy weeks of debate, three paragraphs reflecting concern about the arms race in outer space were finally adopted. The language was a partial victory for the United States, for it wanted to differentiate between military uses of space, which include activities such as reconnaissance, communications, and navigation, and the arms race in outer space, which it perceived to be the cause of the international concern about military space activities.

Other controversial topics discussed at UNISPACE'82 mirrored debate within COPUOS over the past decade. The dispute over whether a country should have the right of prior consent over which television signals are broadcast to it by direct-broadcasting satellites (DBS) was not resolved at UNISPACE'82, although during the U.N. General Assembly meeting in the fall of 1982, a resolution was passed that adopted draft principles which had been developed in COPUOS but opposed by many Western nations (including the United States). The resolution has the status of a nonbinding recommendation so its impact is unclear.

The question of whether a country should have prior consent over the distribution of satellite remote sensing data concerning its natural resources also was not resolved at UNISPACE'82 and remains under discussion within COPUOS. As in the case of DBS, the United States, based on its doctrine regarding the free flow of information, opposes restrictions on the distribution of data.

The issue of guaranteed access to geostationary orbit (GEO) also was highly controversial at UNISPACE'82. A satellite placed in GEO, located 35,900 kilometers above the equator, maintains a fixed position relative to any point on Earth and is therefore quite useful for communication satellites. Developing countries are concerned that developed nations will use all the "good" orbital slots and frequencies in GEO, so that none will be available when they want to launch such systems. Thus, the developing countries want a priori plans to be established through which every country would be guaranteed a place in GEO, whether or not it had a plan to utilize it. Developed countries generally oppose such planning because it would deny access to orbital slots and frequencies that might never be used.

Guaranteed access to GEO is primarily debated under the auspices of the International Telecommunication Union (ITU), a specialized agency of the United Nations. An *a priori* scheme was negotiated for two-thirds of the world in 1977 for direct-broadcast satellites operating at the 12 Gigahertz frequency range. Discussion at UNISPACE'82 could be considered a harbinger of the positions likely to be taken at future meetings; they were much as expected, although the rhetoric of the developing countries seemed a bit more strident than had been anticipated.

Future of International Cooperation

The future of international cooperative space programs looks bright if for no other reason than budgetary situations throughout the world. Nations are becoming increasingly intrigued with the possibilities of the space frontier for commercial applications and scientific endeavors. With cooperative ventures, more can be accomplished.

Despite the International Solar Polar Mission incident, ESA is still interested in cooperating with the United States, and even with the current state of affairs between the United States and the Soviet Union, it can be hoped that tensions will ease and new cooperative endeavors can be started in five to ten years. Japan has shown considerable willingness to cooperate with the United States, and has announced plans to select an astronaut to fly on the Space Shuttle, as has Australia. Other countries may also have a chance to fly their own astronauts on the shuttle—President Reagan made such an offer to Brazil at the end of 1982. The Soviets are also flying cosmonauts from foreign countries (ten have flown already, from Bulgaria, Cuba, Czechoslovakia, East Germany, France, Hungary, Mongolia, Poland, Romania, and Vietnam). Space is truly becoming internationalized.

The biggest problem for the United States in future international space ventures is the annual budgeting process in which the Office of Management and Budget and Congress make yearly decisions on whether or not to continue a program, rather than approving a project for its lifetime. It is not likely that the budgeting process will change, which raises questions as to how the United States can prove that it is a reliable international partner.

In terms of international space policy, the future appears quite bleak. Issues such as prior consent for direct-broadcast satellites and remote sensing, and access to GEO have been debated for more than a decade, and there is little hope that an amiable resolution will ever be reached. Thus, the viability of the international structure for resolving these disputes is being questioned, although in all fairness, the utility of the U.N. system as a whole is currently a matter of great concern to the Reagan administration, not just those components that deal with space. For example, the United States threatened to pull out of the ITU and possibly the United Nations itself because of a highly politicized ITU meeting in 1982. In this case, the ITU, which is theoretically a technical organization for allocating the radio frequency spectrum and orbital slots, degenerated into a political free-for-all when some countries attempted to expel Israel from the organization because of its activities in Lebanon. The issue had nothing to do with the agenda of the meeting but brought to a halt all work of the conference for most of its six weeks. The future of U.S. participation in the ITU remains in doubt.

Overall, though, bilateral and multilateral space cooperation should thrive, but it should be mentioned that competition will also increase. The ESA Ariane launch vehicle, despite two failures out of its first five launches, poses a real competitive threat to the U.S. Space Shuttle. France and Japan have developed advanced technology for communications satellites, and France is also planning to launch a remote sensing satellite that will compete with the U.S. Landsat. Thus, while cooperation has been a major component of the U.S. space program for the past 25 years and will remain so for the indefinite future, competition is likely to play an increasingly important role in setting the stage for the long-term exploration and exploitation of space.

Note

The views expressed here do not necessarily reflect those of the Congressional Research Service. This chapter was written in 1982 and does not take account of developments since then. Copyright 1985 by Marcia S. Smith.

References

Science Policy Research Division, Congressional Research Service, Library of Congress. *World-Wide Space Activities.* Prepared for the Committee on Science and Technology, U.S. House of Representatives. Washington, D.C.: Government Printing Office, 1977.

————. *United States Civilian Space Programs, 1958–1978.* Prepared for the Committee on Science and Technology, U.S. House of Representatives. Washington, D.C.: Government Printing Office, 1981.

Sheldon, Charles S., II, and Marcia S. Smith. *Space Activities of the United States, Soviet Union, and Other Launching Countries/Organizations.* CRS Report 82-45 SP. Washington, D.C.: Library of Congress, Congressional Research Service, February 1982.

Smith, Marcia S. "The First Quarter-Century of Spaceflight." *Futures* (October 1982): 353–73.

National Aeronautics and Space Administration. *International Program Summary.* Washington, D.C.: NASA, published annually.

8

The Proliferation of Communications Satellites: Gold Rush in the Clarke Orbit

Joseph N. Pelton

Introduction

It is remarkable how consistently innovations and "miracle breakthroughs" give rise to new and unsuspected social, economic, and political problems. The use of fire is the ancestor of today's energy problems. Urban renewal problems would never have happened without the invention of farming, which made possible permanent human settlements. At the turn of the twentieth century, the members of the London City Council were agonizing over a major concern that they feared would bankrupt their city treasuries, namely, how (short of financial disaster) hundreds of tons of horse manure might be removed from the streets. The introduction of the horseless carriage was, in its day, greeted as an ecological and financial godsend for the keepers of the city's purse. The automobile, like hundreds of major technological developments before and after, has stimulated progress while at the same time giving rise to new and unexpected social issues and problems. And remarkable as the invention of the geosynchronous communications satellite has proven to be, it likewise has given rise to policy issues that hardly could have been anticipated.

In 1945, Arthur Clarke wrote in detail about the physics of the geosynchronous orbit. He noted that at a distance of some 22,300 miles (35,900 km) out in space, gravity, orbital momentum, and the once-a-day circuit of the Earth all matched! Thus he argued this orbit would be exquisitely convenient for creating a complete global satellite system with only three satellites. His article, published in "Wireless World," was greeted with balanced measures of derision, amusement, and skepticism. Indeed, Clarke (whom I visited in Sri Lanka in 1981) explained to me that he did not bother to attempt to patent the idea at that time, because he did not anticipate the invention of the transistor, which allowed the size of communications satellites to shrink to reasonable sizes and to be unmanned.

Clarke has thus seen, during his lifetime, not only the launching of geo-synchronous communications satellites but also their growth into highly complex and sophisticated systems that carry, at very modest costs, over two-thirds of the world's telecommunications traffic.

Just as Clarke did not foresee the extremely rapid evolution of communications satellites within his lifetime, political and technical leaders who devised the initial international and national policies and institutions for communications satellites did not foresee the wide range of policy issues that communications satellites would generate. This lack of foresight is not really surprising. Experts have shown at various times in history "how an airplane could never fly" or, if it did, have demonstrated "how wind resistance would dictate that it would travel slower than a locomotive." Another scion of science predicted in the 1940s that two computers would handle all requirements of the United Kingdom and perhaps as many as five would be needed for the United States.

It is clearly difficult to project the future and visualize the development of new industries, but once established, dependencies quickly develop for new services and products. Our modern sprawling cities are totally dependent on modern transportation and communication systems. Although automobiles, chemicals, and computers now give rise to "problems," we cannot just stop using them—at least not without causing serious disruptions to society, and perhaps even risking severe damage to modern civilization. Abandonment of modern technologies is, in a practical sense, nearly impossible. This "impossibility factor" certainly applies to our global talking machines—the communications satellites. Pull the plug on the emerging network of domestic, regional, and international satellite communications systems that exist around the world and one is likely to pull the plug on the global economy. Airline travel, commodity trading, banking, television news, oil and energy industries, indeed, international diplomacy and more are heavily dependent on satellite communications. As just one example, it has been estimated that perhaps $3.5 trillion Eurodollar and Asiadollar electronic fund transfers (the equivalent of one-third of the world's annual economic product) takes place over the International Telecommunications Satellite Organization (INTELSAT) global satellite system each year.

Given the size of the satellite communications industry, it is remarkable that it really all began less than 20 years ago. The entire industry, including satellites, earth stations, launch services, insurance, tracking, telemetry, and command operations, plus the value of the services they carry, will probably near $100 billion by the late 1980s, with perhaps 90 percent of this involving services and 10 percent representing equipment and facilities.

In April 1965, the world's first transoceanic satellite system, known as INTELSAT, was created (See Table 1). At the end of its first year of operation, it had in service only 75 full-time voice circuits. In those days, all telephone service on the satellite had to be suspended to allow a medium-quality black-and-white television channel to be transmitted across the Atlantic Ocean. It was from these modest beginnings that the phrase "live via satellite" became a part of our contemporary culture. By 1969 three INTELSAT III satellites (positioned over the Atlantic, the Pacific, and the Indian Ocean regions)

MILESTONES IN INTELSAT'S HISTORY

*U.N. Resolution (XVI) 1721	1960
*COMSAT Act passed in U.S.	1962
*Signing of INTELSAT Interim Agreements	August 1964
*Launch of Early Bird	April 1965
*Achievement of Global Satellite Network	1969
*Definitive Arrangements Negotiations	1969-1971
*Entry into Force of Definitive Agreements	February 1973
•Secretary General assumes office	September 1973
•Management services contract signed	August 1974
•Structure of permanent management arrangements for INTELSAT Approved by Second Assembly of Parties	October 1976
•Director General assumes office	December 1976
•Management services contract amended to reflect new management arrangement	December 1976
•INTELSAT membership reaches 100	November 1977
•Executive organ structure under permanent management arrangements established	1977 through Early 1979
•Technical services contracts enter into effect	January 1979

allowed some 500 million people the world over to see the first moon landing live. From the embryonic satellite system inaugurated by Early Bird in 1965, a vast global network, including scores of domestic and special-purpose communications satellite systems, has now emerged. These systems are almost unrecognizable, in terms of scope and complexity, when compared to the simple devices of the early days.

The INTELSAT System now carries virtually all live international transoceanic television broadcasts, as well as 60,000 full-time telephone channels, on over 1,100 preassigned pathways, to interconnect over 160 countries, territories, and independent possessions around the globe. Furthermore, domestic communications are provided by INTELSAT, under long-term space-segment leases, to more than 20 countries. This service involves more than 40 countries. Separate domestic and regional satellite systems now in operation or planned number more than 50. There is an astonishing array of about 120 communications satellite systems inhabiting geosynchronous orbit. Yet, this proliferation of communications satellites and the growing dependence on satellite communications is invisible at the consumer level and would be noticeable only if these vital services were to disappear.

Emerging Policy Issues

The communications satellites of the 1980s, unlike their forebears of the late 1950s and the early 1960s, are highly complex technological devices that are designed, built, and typically launched into that special orbit known as the geosynchronous orbit (or the Clarke orbit), where they can continuously provide a wide range of communications services to users who access the satellites through ground antennas. These multibillion-dollar satellite services (which include telephone, telex, facsimile, electronic mail, television, videoconferencing, computer networking, and data communications) are now vital means of global and national communications. They are more cost effective and versatile than past ways of communicating.

Given the above, you may ask what could possibly be the problem. Unfortunately, the answer is plenty. First, there is the issue of transborder information flow. Communications satellites are able to relay information across international boundaries, which gives rise to concerns such as privacy of information, intrusion of foreign interests into domestic economies, and unwanted political, cultural, commercial, or religious messages being beamed into a country without the cooperation or permission of its local government. Particularly in United Nations and UNESCO debates, these issues have been given increasing prominence; serious rifts both between "North" and "South," "East" and "West" have occurred. Many nations are moving to enact laws to protect their "domestic industries" in the transborder data flow debate. (It is perhaps ironic to note that the International Telegraph Union, now known as the International Telecommunication Union, was created in 1865 to cope with this issue of transborder data flow in the context of the telegraph).

A second major issue relates to international equity and equality of access to communications satellite services. Because deployment of communications satellites requires a sophisticated technological base and considerable financial resources, many Third World countries have become very concerned about the increasing number of geosynchronous communications satellites being deployed by the developed countries. Their concerns include the fear that by the time they want to deploy or utilize their own communications satellite system, there will be insufficient locations remaining in geosynchronous orbit for their own purposes. These issues came to a head within the International Telecommunication Union's World Administrative Radio Conferences (WARC) held in Geneva in 1979, and were most pointedly expressed in terms of opposition to the status quo principle of first-come, first-served allocation of orbital parking spaces.

A third issue relates to competition. There is today competition between technological media such as satellites and submarine cables and thus between aerospace and cable manufacturers. Likewise, there is emerging competition between domestic, regional, and international satellite systems. Competition has been further promoted by the U.S. government's attempt to deregulate, commercialize, and diversify the telecommunications market.

The fourth issue is an emerging rivalry among the different types of satellite communications service requirements, such as fixed satellite communications services versus broadcast satellite services, versus mobile satellite communications services, versus radio location (or radar) services.

Some have suggested that the use of satellites for communications should ultimately be reserved for those applications that best serve their unique characteristics, such as multipoint-to-multipoint broadcasting applications or mobile services. Others strenuously argue that satellite communication facilities should continue to provide all forms of services, with demand and marketplace considerations being primary. The resolution of these questions will largely depend on which radio frequencies are made available for satellite communications within the framework of the ITU international negotiations and decision-making process over the next five years. Specific technical debates will focus on: (1) the timing of when, as well as which, higher radio frequencies are to be utilized for satellite communications; (2) whether certain applications or geographic areas should be given prime access to the lower and more desirable radio frequencies; and (3) whether accommodation and agreement can be reached between developed and developing countries to meet all technical, operational, and service needs.

The issue of frequency allocation and use within the ITU forum involves not only communications satellites but other terrestrial communications applications as well. There is frequently a dispute over whether terrestrially based use of the lower radio frequencies should be primarily for telecommunications services (a view usually expressed by developing countries) or primarily for broadcasting and mobile communications services (a view usually expressed by developed countries). There are other technology-related issues, such as the continued use of "conventional" dedicated communication satellites (that have been in service for the past 20 years) versus the concept of increasingly complex and large multipurpose space platforms. Some feel that consolidation of a large number of service requirements and a range of different radio frequencies into a single very large spacecraft is a much more efficient use of the geosynchronous orbit than a proliferation of "conventional" communications satellites. The above issues, which of course involve many key technical aspects, also involve certain conflicting industrial and economic interests, particularly in terms of U.S., Japanese, and European aerospace, communications, electronics, and cable industries.

Although this is hardly a comprehensive list of issues that have arisen from the successful development of the communications satellite field over the past several decades, they are indicative of the range of economic, political, and social issues that confront the field today. Since we cannot explore all of the issues in depth, let us focus on what is probably the primary issue that the ITU, INTELSAT, and the communications satellite industries now face, particularly in the context of the World Administrative Radio Conferences scheduled for 1985 and 1987: How can equitable and beneficial access to the geosynchronous orbital arc be assured to all countries, be they less developed, industrializing, or highly developed information societies and, in particular, should such access be based on opportunity, resources, or need?

Into the Maze of Telecommunications Diplomacy

To understand the dimensions of the issue of and assignment access to the geosynchronous orbital arc, one must know more about the major institutions involved. These include: (1) the *International Telecommunication Union (ITU)*, which is responsible for international regulations governing the use of the world's electromagnetic spectrum for communication and scientific purposes; (2) the *International Frequency Registration Board (IFRB)*, which is the unit within the complex ITU structure that maintains the master register of frequencies and records coordination agreements between satellite systems; (3) the *International Committee on Telegraph and Telephone* and the *International Committee on Radio*, (respectively, in the French abbreviations, the

CCITT and the CCIR), which develop internationally agreed-upon recommendations for standards of communications in the field of electronic telecommunications, radio, telegraphy, and television; (4) the *International Telecommunications Satellite Organization (INTELSAT)*, which, operating under two international agreements, now provides two-thirds of the world's overseas telecommunications services among and between some 160 countries, territories, and independent possessions, and also provides domestic communications services to over 20 countries; (5) the *International Maritime Satellite Organization (INMARSAT)*, the agency created under two international agreements for the purpose of providing international maritime satellite services; (6) *UNESCO*, the specialized U.N. agency that is responsible for communications at the level of mass media, journalism, and culture, but that recently, through a new subsidiary the International Programme for Development of Communications—is increasingly involving itself in transborder data flow and telecommunications development issues.

There are also a host of other organizational entities involved in the provision of satellite communications service facilities. Today thousands of people are involved in the very complicated field of international diplomacy on satellite communications.

In theory, the International Telecommunication Union holds every 20 years, among its entire membership, the World Administrative Radio Conference (WARC) to review and revise the worldwide Radio Regulations, including the table of radio frequencies allocations. The ITU Radio Regulations are a vast complex of recommendations, tables, and charts sprinkled, measles-like, with a plethora of highly technical explanatory footnotes. This forbidding document and masses of CCIR and CCITT documents are the essence of the worldwide regulatory process for satellite communications, as well as all other modern forms of communications. They are of the utmost importance not only to the multibillion-dollar satellite communications industry but indirectly to an increasing portion of the world economy.

Given that the telecommunications field is evolving very rapidly in terms of technology and also in its political and social dimensions, the practice has emerged of convening many international conferences to deal with relevant subjects at intervals much more frequent than the traditional 20 years. In fact, in recent years there have been specialized administrative radio conferences on satellite broadcasting, medium-frequency broadcasting, maritime communications, etc.—rather like "kissing cousins" to a full-scale WARC.

In addition to dividing the WARC process into sectors related to services, technologies, or radio frequency ranges (or indeed a combination of these), the ITU has found it useful to divide the world into three regions: Region 1—Europe, Africa, and the Middle East; Region 2—the Americas; and Region 3—Asia and Australia. Regional conferences held to deal with subjects relating to one or two specific issues are called Regional Administrative Radio Conferences (RARCs). In 1983 the RARC for Region 2 was devoted to the allocation of radio frequencies for broadcasting satellite services.

Thus, there is the following labyrinthian maze of units within the ITU structure with which to contend: the intricacies of the CCIR and CCITT, with nearly 50 working groups; the WARC's own Special Preparatory Meetings; the Specialized Administrative Radio Conferences, plus the RARCs; the plenipotentiary meetings of the ITU; and the meetings of the ITU Administrative Council. Seldom does a week go by when an ITU meeting is not held somewhere. On the one hand, this extremely elaborate and complex coordinative process is a remarkable exercise in international cooperation at the scientific and technological levels. It has been said that one form of tyranny is to impute simplicity where it does not exist. Complexity is a necessary response to the high-tech, rapidly evolving field of telecommunications. On the other hand, the need to have a high level of expertise to participate in ITU meetings is in itself a major concern of developing countries that lack both the expertise and financial wherewithal to participate equally alongside the more advanced countries such as the United States, the Federal Republic of Germany, France, or Japan.

The ITU, which is the world's oldest universal-membership intergovernmental organization, has been swept up in the winds of change in the last decade and a half. First established as the International Telegraph Union in 1865, it was essentially designed to coordinate European telegraphy. In 1932, the ITU was reconstituted as the International Telecommunication Union, with a broader mandate to cover all elements of telecommunications and particularly to promulgate regulations for the practical and scientific use of the electromagnetic spectrum. As late as 1945, ITU membership was just over 70 countries, largely dominated by Western industrialized countries. By 1984 the rapid emergence of new countries had pushed membership to over 150. The ITU is the world's largest specialized agency in the United Nations system, and has more members than the United Nations itself. Because the ITU operates on a one-vote-per-country basis, the historical dominance of Western industrialized countries has dramatically changed at a time when the level of technical complexity has greatly increased. Further, the global view of telecommunications has shifted, with modern communications being increasingly recognized as a primary strategic and economic force.

As noted earlier, the last major World Administrative Radio Conference was held in Geneva in 1979; the Radio Regulations of the ITU were then reviewed and extensively revised. Certain unresolved issues led to adoption of Resolution BP-1, which provided for two conferences (now scheduled for 1985 and 1987) to determine the means whereby all countries of the world could be guaranteed equitable access to the geosynchronous orbital arc for communication satellite purposes. Preparations for the conferences are under way around the world. In the United States, for instance, an advisory committee of over a hundred experts and a network of subcommittees are developing positions, strategies, and evaluations of options. U.S. governmental staff (particularly of the FCC, NASA, NTIA, and DoD) and consulting firms are engaged in technical studies, and the State Department is undertaking preliminary soundings around the world. Although the U.S. preparations are perhaps the most complex, a somewhat comparable level of activity has also begun in most OECD countries. Two examples of the level of detail involved are the following: (1) in Japan, the National Space Development Agency is developing technical and legal proposals concerning procedures for removing retired communication satellites from geosynchronous orbit; and (2) the United Kingdom has undertaken a sophisticated study of the merits of using lasers instead of microwave for space-to-earth energy relay.

The Pivotal Issue

Although there are hundreds of technical subjects to be evaluated, discussed, and agreed upon at upcoming WARC conferences, the basic policy issues are few. Here, in my view, are three pivotal issues that, perhaps surprisingly, are heavily dependent upon technical reasoning and studies.

Guaranteed Access to the Geosynchronous Orbit

Developing countries (or "The Group of 77"—which includes most developing, Third World, and "nonaligned" countries and which is now closer to 90) will likely press for a standardized planning formula to divide "orbital arc resources" equally among all countries, regardless of need or ability to soon use the orbit. the idea is to protect the resources for the future. Alternative proposals, some of them quite innovative in nature, will come from developed countries, and will include some of the following:

1. to divide the orbit into broad geographical regional service sectors aimed at accommodating regional demands;
2. to establish procedures to temporarily continue first-come, first-served principles, with penalties for overstaying allocation periods;
3. various economic, technical, and operational plans to "guarantee" access to the orbital arc for Third World countries either by (a) tighter spacing of satellites to accommodate newcomers or (b) firm commitments that multilateral–multiuser satellite systems like INTELSAT will furnish space segment facilities and services adequate to meet communication needs. This latter plan could be effectuated either through long-term lease of space-segment capacity or by means of dedicated space segment.

In fact, the final solution may be a combination of these proposals, since a continued INTELSAT policy of leasing capacity for domestic service can be anticipated even in the absence of the WARC negotiations.

Defense and Governmental versus Civilian or Commercial Applications in Space

Radio location (radar) and, to a lesser extent, military space communications are applications where exclusive use of frequencies is desirable. The parallel growth of defense, commercial, and civilian space communications during the 1970s and 1980s increasingly means that these two types of services (i.e. military and civilian) are in direct competition for radio allocations in the geosynchronous orbital arc. The escalation of this battle, particularly when viewed as a "zero sum game," translates into a difficult and troublesome series of issues to resolve, with a minimum of positive answers in sight.

Conventional "Small," Dedicated Communication Satellites versus Complex, Multifrequency Space Platforms or Clusters

The demands on the geosynchronous orbital arc can be resolved in only a few ways: (1) limitation on growth of services; (2) tighter spacing of satellites, with more interference; (3) more efficient modulation, coding, and processing techniques to allow higher throughput through the same satellite, or higher performance spacecraft, (such as large multibeam orbital antenna farms or satellite clusters) to allow manifold frequency reuse and multipurpose services at a single point on the geosynchronous orbital arc. It is of course possible, indeed likely, that all of the above could happen simultaneously. Philosophically, however, the basic issue is the balkanization of space by the proliferation of dedicated satellites versus integration of space applications into multipurpose and multiuser programs. Today, the trend is toward continued satellite proliferation, with each country, and indeed each company or institutional entity, trying to plant its flag or corporate logo in Clarke orbit. To state the extreme, if each communication satellite system identified for operation by 1985 is used as a trend line for projecting the situation into 2025, we could project some 20,000 separate satellite systems. Although this is of course clearly an impossibility, with only some 120 to 180 orbital slots available in the Clarke orbit, the planners of satellite services have yet to face the need to consolidate domestic services in such a manner as INTELSAT has done internationally.

Prospects for the Future

Despite the complexity of the issues involved and the many technical, operational, financial, and industrial issues that the use of the geosynchronous orbital arc has generated in the past 20 years, there is no reason for despair. Indeed there is a sound basis for optimism regarding the future for the following reasons:

1. *Strong motivation toward research and development.* The challenge of fiber optics, the demand for new satellite service, the growing revenues from satellite communications, and the tremendous demands upon the Clarke orbit all stimulate a strong research and development effort to provide new and more cost-effective technologies. Progress in the field since 1965 has been enormous—a result largely fueled by research and development. The prospects for the future are equally bright.
2. *Technological solutions on the horizon.* Some technical solutions that can multiply the effective use of the Clarke orbit are almost within reach. These include: on-board signal processing; large, multibeam antennas to allow extensive frequency reuse; space platforms; satellite clusters; and intersatellite links. These technologies should allow a more than hundredfold increase in the capacity of the world's satellite systems by the twenty-first century.
3. *Workable institutional arrangements.* The institutional arrangements within which decisions concerning satellite communications must be made are complex and intimidating to the uninitiated. They are, nevertheless, quite workable. Although improvements might be possible with regard to the structure and operation of the ITU, IFRB, CCIR, CCITT, INTELSAT, INMARSAT, and IPDC, by and large these institutions are coping with the significant telecommunications issues with considerable skill and success. The world's telephone and data network is today one of the largest and most smoothly functioning "machines" in the world, while the implementation of the so-called Integrated Services Digital Network (ISDN) in the later 1980s will make this even more the case. This amazing achievement is, in part, a testament to the institutional arrangements in this field.
4. *Effective leadership.* Knowledge, competency, and capable administration, rather than political charisma, remain the key assets of leadership, at least in the field of international telecommunications. The ITU has been fortunate to have had wise and far-seeing leadership.
5. *Spirit of international technical cooperation.* The U.N. systems at times seemed to have fallen victim to petty politics, bloc voting, and endless unresolved debate. It can be confidently said, however, that the ITU has, for the most part, escaped these pitfalls. A telecommunications tradition of over one hundred years remains strong, wherein issues are generally objectively examined on their technical merits. Peer pressure among scientists and engineers to exclude polemics and to seek sensible and equitable solutions has served telecommunications the world over quite well. One hopes the tradition will continue.

As we reach to outer space to seek better ways to serve the people of our small planet, we also see the extent to which we are committed to a common goal and a common fate. Our Earth is only a 6-sextillion-metric-ton mudball—a small planet, in a small solar system, on the outreaches of a medium-sized galaxy, in a universe studded with hundreds of millions of other galaxies. As we gain this cosmic perspective, the desirability of finding global solutions for not only telecommunications issues but for all our fields of endeavor will become increasingly clear. Certainly, in the field of satellite communications, the sky is *not* the limit.

Note

The views expressed in this article are those of the author and are not intended to reflect those of INTELSAT.

References

Jonathan F. Galloway. *The Politics and Technology of Satellite Communications.* Lexington, Mass.: Lexington Books, 1972.

Judith T. Kildow. *INTELSAT: Policy-Maker's Dilemma.* Lexington, Mass.: D.C. Heath & Co., 1973.

Michael Kinsley. *Outer Space and Inner Sanctums.* New York: John Wiley & Sons, 1976.

William E. Lee. *The Communications Satellite Act of 1962: The Creation of New Communications Policy.* Madison: University of Wisconsin Press, 1977.

Brenda Maddox. *Beyond Babel: New Directions in Communications.* New York: Simon & Schuster, 1972.

James Martin. *Communications Satellites.* Englewood Cliffs, N.J.: Prentice-Hall, 1977.

Hamid Mowlana. *International Communications.* Dubuque, Iowa: Kendall/Hunt, 1971.

Joseph N. Pelton. *Global Communications Satellite Policy: INTELSAT, Politics and Functionalism.* Mt. Airy, Md.: Lomond Systems, Inc., 1974.

———. *Global Talk.* Leyden: A. W. Sijthoff, 1981. Distributed in United States by Kluwer Boston, Inc.

———, and Marcellus S. Snow, eds. *Economic and Policy Problems in Satellite Communications.* New York: Praeger, 1977.

Kathryn M. Queeney. *Direct Broadcast Satellites and the United Nations.* The Netherlands: Sijthoff & Noordhoff, 1978.

Delbert D. Smith. *Communications Via Satellite: A Vision in Retrospect.* Reading, Mass.: A. W. Sijthoff International, 1977.

———. *Teleservices Via Satellite—Experiments and Future Perspectives.* Reading, Mass.: A. W. Sijthoff International, 1979.

———. *Space Stations, International Law and Policy.* Boulder, Colo.: Westview Press, 1979.

Marcellus S. Snow. *International Commercial Satellite Communications.* New York: Praeger, 1976.

9

Underdevelopment via Satellite: The Interests of the German Space Industry in Developing Countries and Their Consequences

Jürgen Häusler and *Georg Simonis*

Technological changes in the central capitalistic, industrialized countries not only determine their relationships with underdeveloped countries but also deeply influence the latter's economic, social, and political change. Thus, every technological "revolution" in the center has changed the form in which the periphery was integrated into the capitalistic world system. We believe that only a few nations and peoples have benefited from technological innovations and the consequent modifications in the international division of labor. Most countries are helplessly facing changing production technologies and conditions of capital accumulation because they do not have the political, economic, and cultural competence to adapt the invading innovations to their respective national production and reproduction processes. The countries in the capitalistic periphery were unable to use the inputs from the developed countries to satisfy their national needs. Thus, in our analysis, as the underdeveloped countries failed to adapt, the process of underdevelopment continues and misery increases.

We seem to be entering a new phase in the process of underdevelopment. An immense innovative push is taking place in the present world economic crisis. This push is going to change the conditions of production worldwide. It will create new growth industries and industrial complexes. The integration of telecommunications and data processing (telematics) represents one of the most dynamic sectors of innovation.[1] Satellites are an indispensable part in the worldwide application of this technology to produce, process, store, transmit, and distribute information. The use of satellites provides global networks for communication media, data transmission, mass communication, and the observation and surveillance of the Earth, weather, and atmosphere.

The creation of national, regional, and international satellite systems can be observed in developing countries. It is going to change their relationships to the industrialized countries and profoundly influence their social and political development. It is questionable, however, if the foreseeable technological change is going to improve the living conditions of the mass of the people in the periphery, as the apologists of the new technology maintain. Currently at least, more evidence supports the argument that the new phase of penetrating the periphery with telematic, satellite communication, and remote sensing will bring more dependence, alienation, and exploitation to most underdeveloped countries.

If negative effects are to be minimized, some possible applications of satellites should not be used at all and technology has to be carefully integrated into the national or regional process of mobilizing development potentials. Both suggestions are rather unlikely to be followed. We believe that demand for satellite systems will result instead in applications and forms that will be to the disadvantage of the masses in underdeveloped countries and for most developing countries as well.

The Market for Satellite Systems

The *first* big area in which satellites were used for civil purposes was created in 1964 with the foundation of INTELSAT. The organization offers satellite services for international telephone traffic and the transmission of TV broadcasts. Today, about 150 countries with over 300 Earth stations are part of the INTELSAT network. By the end of 1979, 39 percent of INTELSAT's international telephone lines were used in developing countries.[2]

To satisfy the enormously growing demand for international transmitting capacity, better-performing satellites were developed. So far, all systems for INTELSAT have been developed in the United States. European companies participated in the program to a greater degree for the first time in the production of INTELSAT V, the satellites built by Ford. However, in the production of INTELSAT VI satellites, no great participation by European countries seems to be provided for; U.S. companies, led by Hughes, are dominating the market.

INTELSAT is leasing free transponder capacities to support the creation of national satellite communication systems and has thus opened up the *second* field of application for satellites in developing countries. Via satellites, remote areas can be connected to national telex and telephone networks and can also receive radio and television broadcasts. There is no need to build up expensive cable systems and broadcasting stations anymore. An overview of developing countries using INTELSAT to install a national satellite system is given in Table 1.[3]

Some countries or regions have bought or are planning to buy satellites to build up their national and regional telecommunications systems, as Table 2[4] shows:

TABLE 1
Developing Countries Using Domestic Satellite Services

Country	Start of Service
Algeria	17 February 1975
Brazil	1 July 1975
Chile	1 October 1977
Columbia	1 July 1978
India	1 April 1979
Malaysia	1 August 1975
Mexico	Early 1980
Nigeria	31 December 1975
Oman	15 November 1978
Peru	15 December 1978
Philippines	Mid-1980 (Palapa A and B lease)
Saudi Arabia	1 January 1977
Sudan	15 February 1977
Zaire	15 December 1978

TABLE 2
Developing Countries Using National or Regional
Satellite Systems

Country	Start	System	Producer
Indonesia	1976	Palapa	Hughes
ASEAN	1979/80	Palapa	
India	1982	Insat	Ford
18 Arabian			
States	1982	Arabsat	Aerospatial
Columbia	1984	Satcol	(MBB)
Brazil	1984/85	SBTS	
Panaftel	1985	Afrosat	

It is estimated that in developing countries, regional satellite systems for telephone traffic and television broadcasting are going to be the most dynamic sectors in telecommunications. This estimate includes the *third* field of application: the direct-broadcasting television satellites. This type of satellite, heavy and technically complicated, is still being refined.

The U.S. satellite ATS 6 has already demonstrated some fields of application in developing countries (the NASA Project Site in India). It cannot now be foreseen how quickly a market for television satellites is going to grow in developing countries. Numerous technical, economic, organizational, and political obstacles delay a speedy introduction into the market. Financial problems especially lack solutions. Because of these problems and to improve technology further, the industrialized countries are concentrating their efforts on their respective local markets and on export markets in other industrialized countries. Their long-range perspective, however, aims at the markets in developing countries. Thus, Eurosatellite advertises its direct-broadcasting television satellite with the words: "This multipurpose satellite can satisfy the needs of countries like Spain, Saudi Arabia, Argentina, Brazil, China." And the same industrial consortium says of its smaller "middle-class satellite": "It is the goal of this type of satellite to satisfy the different needs of international programs, for example, in Australia, Africa, the Arabian countries and for Colombia."

Finally, the *fourth* important market for satellite technologies in developing countries is represented by remote sensing satellites. Possible applications of this type of satellite in Earth observation, oceanography, and climatology are manifold.

The large economic, ecological, and scientific potential of remote sensing satellites is just beginning to be used. Only a few developing countries— India, Brazil, Indonesia, and China—are participating in the development of the sensor and satellite technologies, but the developing countries are supposed to be the biggest commercial users of remote sensing systems in the future. Thus, the industrialized countries interested in the commercialization of these technologies support the developing countries in building the necessary infrastructure and in utilizing the data delivered by satellites.

An incentive for the interest of the developing nations in remote sensing technologies has been the Landsat program, under which the United States has provided help to set up receiving stations and to analyze data. Several countries have built or are planning national or regional centers to receive and interpret remote sensing data. Remote sensing data technology is still in a stage of testing (with the exception of weather satellites). Commercialization of data has just begun in preoperational systems in the United States (Eros) and in France (Spot Image). But the leading industrialized countries are hoping for big business in selling the data, the hardware to receive the data, and the software to interpret the data. Today, it is still uncertain when worldwide commercial remote sensing satellite systems can be introduced for many financial, political, and organizational problems are yet to be solved.

To utilize the several possible applications of satellites, large investments will be necessary. Satellites, carriers, broadcasting or receiving stations including cables to users, and appliances such as telephones, TV sets, computers, and printers are needed. The costs for the satellite itself are low in relation to the costs for the ground network. In a study for the German Ministry for Research and Technology (BMFT), the German company Dornier found the following division of costs in rural communications systems in Africa: of a total of $720 million, only 18.5 percent went into the space segments while 81.5 percent went into the ground network.[5]

The imbalance between costs convinces us that, first and foremost, it is the electronic and electric industries producing Earth stations, antennas, and appliances that will benefit from the commercialization of satellites. Because the technical layout of satellites sets the standards for Earth stations, close cooperation between the space and electronic industry is mandatory, as can be seen in the Eurosatellite consortium. In addition, because the various satellite systems are not compatible and standardized yet, developing countries have to purchase satellite systems and ground network from the same industrial consortium.

To be able to participate in the future market for Earth stations, developing countries should start to produce this equipment as soon as possible. India and Brazil are positive examples. With their own production facilities, they could force sellers of satellites in the industrialized countries to adapt to the needs and production capacities in developing countries. However, in that most developing countries are not able to produce the ground segments themselves, they should seek cooperation with companies in the industrialized countries. It then becomes obvious: Those companies in the center that simultaneously produce the satellites, and cooperate with companies in developing countries for local production of receiving stations, will financially benefit the most from the evolving market in the Third World.

The Development of the German Space Industry through European Cooperation

Space technologies are still in a development stage, but some will be commercialized in the near future. Our chapter concentrates on the latter, i.e. launchers and applications satellites. European companies, like Eurosatellite and Arianespace, testify to the coming commercialization of some space technologies. The marketing director of Arianespace is confident: "We intend to make money for our shareholders by 1986."[6]

Additionally, space technologies are being described as "key technologies": they are supposed to be future growth industries, with technological and commercial spin-offs in other sectors of the economy. Thus, they are a central part of a strategy to overcome the present structural economic crisis. They are central to the "modernization" of the economy. That is the view of the European Space Agency (ESA), among others, as well: "We Europeans are very sensitive to the fact that space, like computers, is one of the handful of important, growing fields that modern, Western societies have to develop. It's crucial for our progress."[7]

But, even if it is meant to be "progress for all," naturally the world market for space technologies also evolves through competition and cooperation among the industrialized countries, with an uneven distribution of benefits. A hierarchical international division of labor is taking shape in this technological sector too.

The End of U.S. Dominance: Space Shuttle versus Ariane

The year 1981 was important for the U.S. space industry:

- The dream of a reusable space carrier became a reality with the launch of Columbia on 12 April 1981.
- However, on 19 December 1981 ESA sent the European launcher Ariane on its fourth and last preoperational flight.

NASA's dream (to be the first) and nightmare (to lose the top position to Europe or Japan) are running neck and neck in the competition to develop launchers. Comparable to the Sputnik shock in 1957, though with less public notice but probably greater commercial consequences, NASA has to watch U.S. domination in the launcher market deteriorate.

With Ariane, Europe is entering a market that was previously monopolized by the United States. By 1990, the market should be worth $6 billion, with 200 satellite launches.[8] ESA is expecting a third of these orders for Ariane, which seems reasonable when the two competitors are compared:

- A launch with Ariane is only slightly more expensive, if at all, than the Space Shuttle: $24 to 25 million versus $18 to 28 million.
- Ariane seems to be more reliable as far as costs, technology, and starting dates.
- With Ariane (and right now only with Ariane) the geostationary orbit can be reached more easily; this is mandatory for telecommunications satellites.
- Especially important for developing countries are the favorable terms of payment: prefinancing is available ("launch now, pay later") and terms

are tailored to each potential customer. European banks offer low-interest loans to developing countries using Ariane.

Consequently, there exists a fair amount of orders for Ariane. Over 40 commitments (33 orders, 10 options) have been allotted until 1986. In addition, the following entities already use Ariane as their launching vehicle:

- satellites of ESA or other European national satellites (France: Télécom, SPOT; France and Germany: TV-Sat/TDF);
- other satellites of non-European and developing countries (Australia, Canada, Arabsat, Colombia, Brazil, Mexico, India, and South Korea);
- U.S. companies (Western Union, General Telephone and Electronics, Southern Pacific Communications, R.C.A., and A.T.&T.); and the international organizations (Eutelsat, Inmarsat, INTELSAT).

And the future promises to be even more successful, according to *Time*: "Western European space officials can be expected to make a big push for Ariane, especially among developing nations, at Unispace 82."[9] By 1990, Director Iserland of ESA expects 10 to 15 percent of orders for Ariane to come from developing countries.[10]

Commercialization of Ariane (production, selling, marketing, and financing) was transferred to the European space consortium Arianespace in 1980. Shareholders of Arianespace include 14 banks and the élite of European space manufacturers.[11] The 51 shareholders from 11 European countries who founded the first commercial space transport company had a goal; "conquering the space market," and a common understanding; "The time has come to set up space transport activities on an industrial and commercial basis."[12] By 1985, Arianespace wants to produce and sell 4 to 5 rockets annually and make profits that way.

Ariane was developed in the framework of ESA at a cost of $1.2 billion. It and Spacelab were ESA's biggest programs in the 1970s. France has been dominating the planning, development, financing and commercialization of the Ariane project: Ariane was administered by CNRS, built by Aerospatiale, and launched by the French "Guiana Space Centre"; France financed two-thirds of the project (the rest was underwritten by the Federal Republic of Germany, Belgium, Denmark, Italy, the Netherlands, Spain, Sweden, Switzerland, and Great Britain); and French companies own 60 percent of the shares of Arianespace.[13]

Germany was the second biggest contributor to the Ariane program. Its initial skepticism about the project accounted for its lesser role. It financed 20 percent of the development costs, and German companies hold 20 percent of the shares of Arianespace. The German companies ERNO, MAN, MBB,

TABLE 3
Arianespace Shareholders

Countries	%	Companies[1]	%	Banks	%
France	59.25	Centre National d'Etudes Spatiales (CNES)	34.0	Credit Lyonnais	0.5
		Societe Nationale Industrielle	8.5	Societe Financiere Auxiliaire	0.49
		Societe Europeenne de Propulsion		OPFI Paribas	0.4
		(SEP)	8.5	VALORIND	0.4
		MATRA	3.6	Banque Vernes et Commerciale de Paris	0.2
		L'Air Liquide	1.85	Banque Nationale de Paris	0.01
West Germany	19.6	Maschinenfabrik Augsburg-Nurnberg (MAN)	7.9	Dresdner Bank	0.3
		ERNO Raumtechnik	5.2	Bayerische Vereinsbank	0.3
		Messerschmitt-Bolkow-Blohm (MBB)	2.8	Westdeutsche Landesbank	0.3
		Dornier System GmbH.	2.8		
Belgium	4.4	Societe Anonyme Belge de Constructions Aeronautique (SABCA)	2.4		

[1] Companies holding more than 1.6% shares

Source: Arianespace publications.

TABLE 3 (continued)
Arianespace Shareholders

Countries	%	Companies[1]	%	Banks	%
Italy	3.6	SNIA VISCOSA	1.6	Instituo Bancario San Paolo di Torino	0.2
Switzerland	2.7	CONTRAVES AG	2.15	Union des Banques Suisses	0.3
Spain	2.5	Construcciones Aeronauticas SA (CASA)	1.9		
Great Britain	2.4			Midland Bank Ltd.	0.2
Sweden	2.4	Volvo Flygmotor Aktiebuiag	1.6		
Netherlands	2.2	Fokker VFW BV	1.9	Allgemene Bank Nederland NV	0.3
Denmark	0.70			Copenhagen Handelsbank	0.2
Ireland	0.25				

[1] Companies holding more than 1.6% shares

Source: Arianespace publications.

and Dornier participate in the production of Ariane, especially the second stage. At the same time, the Federal Republic has broadened its involvement in international launcher programs, which is characteristic of its industrial policy in the sector of future key technologies. More than other European countries, Germany, especially through the assistance of ERNO and AEG-Telefunken in the development of Spacelab, in participating in the success of the main competitor of Ariane, the U.S. Space Shuttle. It is more interested in the development of the cargo of the launcher Ariane: the applications satellites.

Competition and Cooperation in Europe: Applications Satellites

European applications satellites are mostly developed within the framework of ESA. ESA was founded in 1975 as a merger of ESRO and ELDO "to provide for and to promote, for exclusively peaceful purposes, cooperation among European states in space research and technology, and their space applications, with a view of their being used for scientific purposes and for space applications systems."[14] Nonetheless, whenever commercialization of certain types of satellites seems to be foreseeable, then national interests and competition gain ascendancy over cooperation.

In fulfilling its mission, ESA has reached a critical point because:

- fiscal crisis in all member (and contributing) states (Belgium, France, Great Britain, the Federal Republic of Germany, Denmark, Ireland, Italy, the Netherlands, Spain, Sweden, and Switzerland) will force the ESA budget to decline in real terms after years of automatically increasing budgets:
- the closing and the uncertain future of the Ariane and Spacelab programs, which until recently dominated the ESA budgets, make new long-range planning of projects and reevaluation of present activities necessary; and
- ESA developments result more and more in commercial products (Ariane, TV, telecommunications, and remote sensing satellites), and thus member states are becoming more interested in handling these programs nationally or bilaterally.

Plans of ESA General Director Quistgaard for the 1980s take these tendencies into account.[15] He is suggesting:

- a decrease in the longer term of ESA annual budgets by 25 percent to under $600 million;
- increased spending on scientific missions; and
- the relatively increased importance of preoperational satellite programs.

The applications satellite programs in the past contributed a quarter to a third of ESA budgets. ESA will promote two projects in the future: the development of the remote sensing satellite *ERS-1* and the telecommunications satellite *L-Sat*.[16]

It is clear to us that the United States is losing its dominant and monopolistic position in remote sensing technology to European products, in spite of a huge lead established by the first American remote sensing satellite, Landsat-1, started on 23 July 1972, and in spite of Europe's experience being limited to weather observation (Meteosat-1 and -2, Sirio-2). According to former NASA manager McCandless, in ten years of operating Landsat satellites, the United States did not make any relevant technological progress.[17] Progress was achieved by two European satellites. To compete directly with Landsat,[18] the French space industry (a consortium led by Matra) developed the remote sensing satellite SPOT, gaining a technological lead. "By the radiometric quality of its multispectral imagery, its enhanced spatial resolution, and its oblique viewing capability, SPOT offers new fields of applications to the world community of users."[19] The lead promises to turn into a commercial advantage. Using Arianespace as an example, SPOT Image was founded to commercialize the data after the satellite came into service.

Without a direct competitor (the American Seasat-1, launched in 1978, ceased to function after six months),[20] ESA is developing ERS-1, a satellite for ocean and coastal exploration. The German company Dornier is managing the project scheduled to start in 1988, based on the French SPOT but with one decisive innovation: all-weather microwave instruments. Using this technology, the second ESA remote sensing satellite ERS-2, for earth observation, will be competing with Landsat and SPOT: "Apart from the prime concerns of ERS-1, the presence on board of new instruments with all-weather capability will usefully prepare their future utilization for terrestrial remote sensing."[21]

For decades, the United States, led by Hughes and Ford, dominated the world market for telecommunications satellites. All INTELSAT satellites (including the new series INTELSAT VI) have been built under the leadership of these corporations.[22] The next generation of telecommunications satellites, the direct-broadcasting TV-satellites, will probably also be pioneered by the United States.[23]

The European industry can attempt to gain only a small but increasing portion of the world market for telecommunications satellites. But this attempt might very well be successful because of: (1) European technology that complies with world market standards; (2) a concentration on "open" markets, i.e. in developing countries; and (3) the desire and legal power of European states to have their satellite telecommunications equipment "made in Europe."

The idea of European cooperation has been replaced by the reality of national competition in the development of telecommunications satellites. All

European projects in the ESA framework face national or bilateral competing projects.

Inside ESA, Great Britain is leading the telecommunications program.[24] British Aerospace is the prime contractor for all satellites:[25]

- the experimental OTS (Orbital Test Satellite), launched in 1978;
- the operational ECS (European Communications Satellites) ready to be launched;
- and the maritime application, MARECS (Maritime European Communications Satellite).

France is developing its own national and commercial telecommunications satellite, Télécom. The prime contractor is Matra. But this system will not be operational until long after IBM's Satellite Business System is widely available. The satellite is going to be operated by the French telecommunications company DGT.[26]

The first European direct-broadcast TV satellite will not be launched by ESA. Instead, the bilaterally developed French-German TV-Sat/TDF, a follow-up of the Symphonie project,[27] will be first.[28] The satellites are built under a German-French governmental agreement by Eurosatellite. Costs amount to about $290 million. Participating companies are Germany's AEG-Telefunken and MBB, France's Aerospatiale, and Thomson-CSF, each holding 24 percent of Eurosatellite's shares, plus Belgium's ETCA, with 4 percent. The "European competitor," L-Sat, with program expenditures of $520 million, will be launched at the earliest in 1986.[29] Again, British Aerospace is the prime contractor.

The ESA program, and especially the above-mentioned applications satellite projects, are based on an industrial policy whose "keystone is to foster the development of a competitive European space industry."[30] But even success can hurt. With the help of ESA programs, national corporations can successfully compete on the world market, but in turn ESA loses its importance, and competition replaces European cooperation. The French satellites SPOT and Télécom and the German-French TV-Sat/TDF are examples of this development, as are France's first space program under Mitterand[31] and the discussion of the new fourth German national space program.

The German space industry (Dornier, MBB, ERNO, and DFVLR) published a memorandum,[32] thought of "as a help in the decision-making process and thus a basis for a long-term space policy in the Federal Republic." It calls for "national developments," the consideration of "national interests," and a "resolutely national program." The "Fourth Space Program of the Federal Republic of Germany" (1982) fulfills industry's expectations. Its motto could be summed up: international cooperation as far as necessary (basic research, future launcher systems, remote sensing satellites), national projects as many as possible (telecommunications), and more financial participation from industry itself.[33] Even the last request seems to be realistic. The ministry and industry share the assessment that some space technologies are just short of commercialization. In other words, they expect that private investment in space technologies will soon be profitable. Industry describes itself as being in the 1980s in the third developmental phase: "Demonstration of profitability in certain areas."[34]

The Shibboleth of "Modell Deutschland"—the German economic miracle in which the government and industry worked together in creating and commercializing technologies—is expected to be invoked again to develop key space technology.[35] Requested already by industry and government is specialization in "telecommunications satellites, for example, TV satellites and their commercial introduction in the world market." Specifically mentioned are the TV-Sat/TDF project, "remote sensing satellites, sensors and data used"; Dornier's role in ESA's ERS program; and MBB and its microwave sensor.[36] The German industrial strategy in space seems to have two aspects: *participation* in as many technical developments as possible, and *specialization* in high technologies.

The German Marketing Strategy

All German space programs have had a strong bent toward the world market, which reflects the structure of the German export-oriented economy and the tradition of German economic and research policy.[37] Cooperation between state and industry in applications satellites, especially telecommunications satellites, demonstrates this orientation.

Until now, the Federal Republic has not sold any commercial satellite system to developing countries. Its space industry, however, is participating in the world market as subcontractors on the INTELSAT satellites. German companies contributed about 10 percent to the production volume of nine INTELSAT V satellites.[38] MBB is now in competition with U.S. companies to receive a contract for the Colombian satellite system Satcol; its success would be considered a major breakthrough. The German space industry wants to sell in the Third World, saying, "There is a growing interest in developing countries in the use of space technologies to solve specific national and regional problems. Some Arab, Asian, African, and Latin American countries are especially interested in using telecommunications and remote sensing satellites for national development."[39]

To secure the potential Third World market for its space and electronic industries, the Federal Republic can make use of three instruments:

● Direct financial support of research and development projects that should lead to commercial products.

- Scientific and technological cooperation with developing countries; bilateral agreements have been signed to support developing countries in building the necessary scientific-technical infrastructure.
- Economic cooperation with developing countries; technical and financial aid will be given to specific projects in developing countries.

Currently, the state mainly provides direct financial support to test and demonstration projects in satellite telecommunications; the direct-broadcast TV satellites TV-Sat/TDF are an example. German industry shows a particular interest in this satellite system: "Direct broadcasting TV satellites represent the fastest and most economical way to set up a television network in developing countries. That is a chance for German and French industry to participate in the world market for high technologies."[40] This view is shared by the German government.

The predecessor of the TV-Sat/TDF project was the German-French satellite Symphonie. Its possible applications were tested in over a hundred experiments between 1975 and 1980 in 35 countries. Cooperation with developing countries—among them India, the People's Republic of China, Iran, Egypt, Saudi Arabia, Libya, Tunisia, and the Ivory Coast—was an essential part of the demonstration programs.[41] No long-ranging governmental agreements were signed but the Federal Republic institutionalized scientific-technological cooperation with some semiperipheral or semi-industrialized countries. Bilateral technical and scientific treaties have been ratified with Argentina (1975), Brazil (1971), India (1974), Indonesia (1980), Iran (1978), and Spain (1973).[42] The more than 50 specific projects undertaken in the framework of the treaties show three areas of special emphasis: observation of atmospheric phenomena; applications of telecommunications satellites; and interpretation and use of remote sensing data plus development of sensors.

The emphasis is based on the following strategic assumptions:

- There is not enough launching room for experiments with atmospheric rockets in Germany. By providing launch sites, the semi-industrialized countries can participate in the preparation, managing, and interpreting of experiments.
- The likelihood of receiving orders for operational systems increases with more intensive cooperation with developing countries, including preparations for satellite telecommunications, the instruction of specialists, and the setting up of local production capacities for ground segments. Furthermore, scientific-technical cooperation represents a valuable source of information on planned programs, and thus enables German industry to anticipate future demand.
- To help developing countries build the capability for analyzing and using remote sensing data, and to help them establish infrastructural prerequisites

for the introduction of commercial systems, it is best that they participate in development. Coproduction of sensors paves the way for marketing of sensor systems.

Bilateral technical and scientific treaties have been signed between Germany and semi-industrialized countries. The treaties ask for a considerable amount of research and financial contributions from the developing countries. Space applications projects with developing countries are administered by the Ministry for Economic Cooperation (BMZ) through its "technical and financial cooperation." According to the official German Unispace brochure (August 1982), more and more foreign aid projects include the use of space technologies. Three big projects are particularly noteworthy:

- In a joint effort with the World Meteorological Organization (WMO) in Geneva, the Federal Republic delivers and installs receiving stations for Meteostat data in Egypt, the Sudan, Kenya, and Tanzania.
- A regional center for receiving data from Landsat, SPOT, and Meteosat has been set up in Upper Volta. Eleven countries in West Africa participate in the program that is financed by Canada, the United States, France, and Germany.
- A study prepared by the Federal Republic in cooperation with the International Telecommunication Union (ITU) in Geneva examines possibilities of using telecommunication satellite systems in remote areas of Africa.

The efforts of the Federal Republic to conquer markets for satellite applications in developing countries seem limited when compared with those of the United States, but they are rapidly increasing. A fierce fight over the market for space technologies is on the horizon. We anticipate that the U.S. technological leads will disappear in the next generation of space technology and satellites, including direct-broadcasting satellites and all-weather sensor systems, and that France and Germany will try to position themselves advantageously in the growing markets.

Consequences for the Third World

Satellites will play an important role in the future development of the periphery, for they are an integral part of global information and communication systems and instrumental in observing the Earth, oceans, and atmosphere. Whether used in the interest of ruling elites or the poor masses in developing countries, satellites are going to be massively used and will have inevitable consequences for the societies. How they are going to be used depends on the underlying logic: the logic of technology or the logic of the society.

Centralizing Tendencies

Our research indicates that the political elites in the Third World strongly support the introduction of satellite systems to reinforce their power. National communications systems enable them to continuously gather information about activities in the provinces and to influence the population via the mass media. International communications channels increase the influence of the elites from industrialized countries, open up information leads for those in power rather than for their political opponents, and make cooperation with international partners easier. Remote sensing satellites disclose new resources. Central agencies that analyze and control data get numerous opportunities to affect developments in remote areas.

Production and management of satellite systems result in consolidated techno-structures, for only highly paid experts can install and run the systems. Decentralized use of telecommunications satellites, remote sensing data, home computers, and TV satellites is possible, but the structure of satellite systems is highly centralized. Production of hard- and software is concentrated in a few production sites, and technical control and use of data are likewise brought together. An excellent example of the centralizing tendencies of satellite systems is seen in the use of direct-broadcasting satellites for television programs in remote areas. Television accelerates in the spread of values and behavioral patterns from the capitalistic, industrial societies, thereby homogenizing cultures.

Transnational Integration

Worldwide satellite communication systems are changing the relationship between developed industrialized societies—such as the United States, Japan, and the Federal Republic—and developing countries. New technical means in data processing over long distances, transmission of texts, and video communications threaten to stop attempts of national industrialization in many developing countries and further limit these countries' political and social independence.

- Multinational companies are going to use new means of international communications to organize their production and labor forces more efficiently; decentralization and centralization are parts of the process. Where office work, sales, and manufacturing processes have been within the purviews of subsidiaries, telematics enables companies to perform these functions better and more cheaply from a central location. Consequences for developing countries are twofold: they lose jobs because of technological obsolescence (although other jobs might be created), and branch operations become more fully integrated into parent companies. The local businesses in this way become part of a worldwide production network, and the locus of control shifts to outside the country.

- Most developing countries have to buy hardware and software for their new communication systems from industrialized countries. Their dependence, the outflow of foreign currency, and their foreign debt are consequently going to increase.
- Developing countries must transmit their data to processing and storage centers in the industrialized countries, and then buy the processed data, an arrangement that leaves the door open to external political pressures and financial exploitation.
- International media companies are going to market the educational, entertainment, and advertising programs that developing countries need to run on their national commercial TV. To finance expensive communication systems, the use of international programs will become mandatory; cultural alienation and the adoption of consumer societies' values are consequences. Both processes are traditionally called "development," although they mean underdevelopment for the rural masses.

Cleavage of Society

Centralizing tendencies in developing countries, as well as increased transnational integration, are going to deepen the cleavage in those societies where a small modernized core and marginal underdeveloped areas exist at the same time. Enthusiastic technocrats anticipate that the communications infrastructure in rural and backward areas will result in the homogenizing industrialization of these regions. The opposite seems true, in spite of the intentions of the technocrats. A combination of three processes is responsible for this opposite result:

1. First and foremost, we believe that new communication systems are going to strengthen the centralization, transnational integration, and dependence of developing societies. It is only relatively faster modernization and an extension of already industrialized urban core zones that will be the consequences. Thus, in spite of all attempts to develop rural areas, they will remain as they are.

2. The integration of rural areas into national and international communications networks and the use of new information channels to instruct the rural masses are likely to break up cultural traditions, restructure social systems, and increase the orientation of daily life to values and standards that prevail in the industrialized core. This paves the way for rural emigration. City life becomes attractive when needs originating externally cannot be satisfied locally.

3. New means of communications should facilitate the industrialization of the agricultural sector, thus substituting capital for labor. Satellite-based sys-

tems to control chemical inputs, artificial irrigation, and the ripening process will be used in continuing "green revolution." This capitalization of agriculture, we expect, will drive massive numbers into urban centers and tend to deepen the social and economic cleavage between rural and metropolitan areas as well as the cleavage between modern sectors and slums in the cities themselves.

These three described processes exist to differing degrees in the various countries of the periphery of the capitalistic world system. Their differentiation, another result of technological changes, has not been discussed here. This description tries to point to tendencies that seem inevitable if the present evolution of social and political structures continues. To turn technical progress into an improvement of living conditions of the masses, underdeveloped countries have to loosen their ties with industrialized countries. Only a selective uncoupling from the capitalistic center and the creation of national or regional capacities to judge critically modern technologies distributed by the center will enable underdeveloped countries to adapt technological systems to their needs.

Notes

1. S. Nora and A. Minc, *Die Informatisierung der Gesellschaft* (Frankfurt/Main: Campus, 1979).
2. All figures, Joseph N. Pelton, *Global Talk* (Alphen aan den Rijn, the Netherlands: Sijthoff and Noordhoff, 1981).
3. Ibid., pp. 213–14.
4. International Telecommunication Union, *Nineteenth Report by the International Telecommunication Union on Telecommunication and the Peaceful Uses of Outer Space* (Geneva: ITU, 1980); J. N. Pelton and M. Snow, eds., *Economic and Policy Problems in Satellite Communications* (New York: Praeger, 1977); Pelton, *Global Talk*.
5. W. Kriegl and W. Laufenberg, *Kommunikationssatelliten-system für Africa*, BMFT-FB-W 80-016 (Bonn: BMFT, September 1980); *Nachrichtentechnische Zeitschrift* (Offenbach) (NTZ), April 1984, p. 250 .
6. *Time*, 8 February 1982.
7. Ibid.
8. All figures, publication of European Space Agency (ESA): *Europe's Place in Space* (Paris: 1981), *Space Science and Technology in Europe Today* (Paris: 1982), and *Arianespace* (Paris: n.d.); *Le Monde*, 12 May 1982; *Neue Züricher Zeitung* (NZZ), 21 December 1981, and 20/21 March 1982; *Der Spiegel*, January 1982; *Süddeutsche Zeitung (SZ)*, 21 December 1981, 13 May 1982, and 12 May 1982; *Time Magazine*, 2 August 1982; *Die Zeit*, 19 March 1982.
9. *Time*, 9 February 1982.
10. Personal correspondence.
11. See Table 3.
12. Cited from Arianespace publications.
13. See Table 3.

14. *ESA, Europe's Place in Space.*
15. *Aviation Week and Space Technology (AW&ST)*, 9 March 1981; *Umschau* (Stuttgart), April 1982, p. 379; *Le Monde*, 17 March 1982.
16. For an overview, see Member States of the European Space Agency, *Joint National Paper*, A/Conf. 101/NP/37, Second U. N. Conference on the Exploration and Peaceful Uses of Outer Space, 1981. Cited hereafter as *Joint National Paper.*
17. According to *Die Zeit*, 16 July 1982.
18. See *AW&ST*, 14 July 1980.
19. *Joint National Paper*, p. 23.
20. *Frankfurter Allgemeine Zeitung (FAZ)*, 14 July 1982.
21. *Joint National Paper*, p. 23.
22. See INTELSAT, *INTELSAT Is* . . . (Washington, D.C.: INTELSAT); *New Scientist*, 15 April 1982; *FAZ*, 5 May 1982; *AW&ST*, 22 February 1982; *Umschau*, November 1982.
23. *FAZ*, 13 September 1982.
24. NTZ, April 1982, pp. 210.
25. ESSA, *Europe's Place in Space; New Scientist*, 20 May 1982; *Economist*, 28 February 1981.
26. *Blick durch die Wirtschaft (BDDW)*, 11 August 1982; *Economist*, 10 January 1981.
27. *Symphonie Symposium Berlin 1980, Proceedings* (Berlin, 1980).
28. *AW&ST*, 21 April 1980; *Le Monde*, 19 August 1982.
29. *SZ*, 2/3 January 1982.
30. *ESA, Europe's Place in Space.*
31. *FAZ*, 23 June 1982.
32. Deutsche Forschungs und Versuchsanstadt für Luft und Raunfahrt (DFVLR) et al., *Memorandum zur Zukunft der Raumfahrt in Deutschland* (mimeo, 1982). DFVLR is the German aerospace research establishment (roughly equivalent to NASA research centers).
33. Bundesministerium für Forschung und Technologie (BMFT), *Viertes Weltraumprogramm der Bundesrepublik Deutschland* (Bonn: 1982), p. 5. BMFT stands for the Federal Ministry of Research and Technology.
34. DFVLR et al., *Memorandum zur Zukunft* p. 6; see also BMFT, *Viertes Weltraumprogramm der Bundesrepublik Deutschland*, pp. 10–12.
35. DFVLR et al., *Memorandum zur Zukunft der Raumfahrt in Deutschland*, pp. 9–20.
36. Ibid., p. 23.
37. Frieder Schlupp, "Modell Deutschland and the International Division of Labour: The Federal Republic of Germany in the World Political Economy," in *The Foreign Policy of West Germany*, ed. E. Krippendorff and V. Rittberger (London and Beverly Hills, Calif.: Sage, 1980).
38. DFVLR et al., *Memorandum zur Zukunft der Raumfahrt in Deutschland*, p. 14.
39. BMFT, *Viertes Weltraumprogramm der Bundesrepublik Deutschland*, p. 9.
40. DFVLR et al., *Memorandum zur Zukunft der Raumfahrt in Deutschland.*
41. *Symphonie Symposium Berlin 1980, Proceedings*, pp. 46–47.
42. *Bundestagdrucksache* 8/3652; Arbeitsgemeinschaft der Gross Forchungseinrichtungen, (AGF; Association of National Research Centers of the Federal Republic of Germany), *Scientific and Technological Cooperation with Developing Countries*, (Bonn: AGF, 1979); BMFT, *Viertes Weltraumprogramm der Bundesrepublik Deutschland.*

10

The Controversy over Remote Sensing

Jean-Louis Magdelénat

Introduction

Human capabilities have been enhanced throughout history, and one of the great strides in the journey of progress began in October 1957 with the launching of the first satellite, Sputnik 1, by the Soviet Union. A great deal has been achieved since then and prospective achievements are even greater. Telecommunications and remote sensing are two of the most important Earth-oriented space activities. They not only have influenced the social, economic, and political life of the most industrialized and developed countries, but have already attracted the attention of the whole world and had an impact on the infrastructure of international relations. It seems no longer possible for national communities to hide behind the conventional political boundaries of their states.

Faced with this reality and bearing in mind the divergent social, economic, and political convictions throughout the world that dictate different ends and means on the one hand, and aware of the potential benefits that could be derived from space science and technology not only for the most advanced communities but also for developing and poorer countries on the other hand, states are trying to reach consensus on the important political and legal issues raised by these new developments. The United Nations Committee on the Peaceful Uses of Outer Space (COPUOS) and its scientific and technical and legal subcommittees have been the forum for serious negotiations on such issues as the rescue of astronauts, liability for damage caused by space ac-tivities, direct broadcasting by satellites, and remote sensing. Agreement has been reached on the first issues, but the last are still under discussion.

In this presentation an attempt to shed some light on the controversial issues pertaining to remote sensing will be made, beginning with remote sensing techniques, and, after examination of some examples of existing and planned commercial remote sensing activities, legal aspects of the issues will be discussed. In conclusion we will try to see what prospects the future holds for remote sensing activities and international relations. There are military uses of remote sensing, which this presentation will not discuss, and peaceful uses, mainly observation of the natural resources of the earth, to which this presentation is confined.

Remote Sensing Techniques

Remote sensing is not a new technique. It has been used and still is carried out by other means than satellites, including high-altitude aircraft, conventional aircraft, light aircraft, helicopters, and balloons.[1] Remote sensing by satellite comprises the following basic elements and steps:

1. One or more satellites orbiting the Earth in a manner suitable for the desired ends and goals to be achieved in the remote sensing process, and carrying sensors, including special and sophisticated cameras, mechanical scanners, and side-looking radar.
2. Ground stations for receiving data from satellites and other related equipment.
3. Skilled personnel, including electronic technicians, and other photointerpretation specialists and computer programmers and systems designers.
4. There are two main types of satellite: the first gathers information through sensing solar radiation reflected by the Earth's surface and/or the radiation emitted from the Earth's surface (called the "passive technique"); the second type of satellite does not rely on sensing radiation but emits its own signals and then senses its signals on their return when reflected by the Earth's surface (called the "active technique"). The satellite may also record the sensing and store it, send the sensed information instantaneously to ground stations, or do both of these operations.
5. It is essential that data gathered in the above manner be supported by data collected in the field (called the "ground truth").
6. Information can be released in primary form or can be processed and analyzed, then compared with other information collected from the field according to the goals set out for the remote sensing operation.[2]

Commercial Uses of Remote Sensing by Satellite Techniques

Remote sensing techniques are used nowadays to achieve many ends, including

1. Hydrology in agriculture, to determine: (a) the availability of surface and ground water throughout the year for the purpose of designing storage,

runoff-river irrigation systems, rural and livestock water supplies, and fishery requirements in rivers and lakes; (b) flood magnitudes and frequencies for the design of spillways on barrages and dams, and for flood control and protection works; (c) drainage of excess irrigation water and swamps; (d) water quality for different purposes; and (e) erosion and sedimentation, including estimation of reservoir sedimentation and control works.[3]

2. Planning, development, and management of grazing land; in both macro- and microplanning, as well as in management, it is very important to have information on the land use (e.g. urban, cropland, rangeland, forest, pasture, rock, and water), the type of vegetation, the type of soils, the location of water points, the availability of communications (roads, canals, railways, etc.), the distribution of human and livestock populations, and data about the country's natural resources. Remote sensing by satellites can provide much of the information needed.[4]

3. Mineral exploration, including oil and natural gas; remote sensing by satellites could be used to explore, locate, and determine the quality and feasibility of extracting natural resources.[5]

It is unnecessary to point out the importance of these remote sensing applications not only to the most industrialized countries but also to developing and poorer nations. This importance becomes crucial when considered in the context of present day circumstances and future prospects concerning the worldwide problem of economic welfare and food availability.[6] Developing and poorer countries do, however, need technical assistance and financial support to adopt these new techniques and methods. This is what the United Nations has been trying to do through its various specialized agencies, particularly COPUOS. A practical solution may be found in commercializing remote sensing activities. To explain this point, let us look at two examples: the United States Landsat and the French project SPOT.

United States Landsat

Landsat 1, 2, and 3 were launched in 1972, 1975, and 1978, respectively.[7] Landsat 1 and 2 have already returned to Earth after providing useful information about types and the conditions of the Earth's vegetation, surface, soil, and water resources.[8] A new program of Landsat spacecraft was initiated with the launch of Landsat-D in 1982.[9] The second spacecraft in this series Landsat-D-Prime is planned to be launched in the later 1980s.[10] This program marks a new phase in the improvement of remote sensing by satellite techniques,[11] including the experimental high-performance multispectral instrument called the Thematic Mapper. The Landsat program is increasingly important for observation of the Earth's resources for many purposes, including land use planning.[12]

Landsat has attracted the attention of many nations, particularly the developing and poorer countries like Indonesia, Pakistan, and Bangladesh. It has ten ground stations in foreign countries, with more stations under construction or the subject of negotiation.[13]

SPOT

The French Experimental Earth Observation System (Systéme probatoire d'observation de la terre, or SPOT) was conceived and designed by the Centre national d'études spatiales (CNES) and is being built by France in association with Belgium and Sweden. The initial SPOT satellite will be put into orbit by an Ariane 2 launcher.[14] In 1981 the French government decided to produce the second flight model of the satellite, with the aim of launching it to ensure continuity of user service from 1985 to 1988.[15] Unlike previous Landsat satellites that used a frequency in the 2 GHz band, Landsat-D and SPOT will use a frequency in the 8 GHz bands for the high-resolution channels, which will entail a major modification in the Landsat receiving station facilities that are now using the 2 GHz band. This is the reason that the French CNES and the U.S. National Aeronautics and Space Administration (NASA) have made joint efforts to ensure maximum compatibility for satellite–receiving facilities, including the sharing by both of them of heavy–duty and expensive equipment, such as antennas, receivers, and high-density recorders.[16] Brazilian, Canadian, and Swedish authorities have already decided to acquire stations using an 8 GHz band (Landsat-D/SPOT compatible stations), and studies are also being carried out in East Africa, Australia, India, and Japan with a view to initiating negotiations with the CNES.[17] Entirely new stations are planned in Bangladesh, Upper Volta, and France for handling the reception of SPOT data in addition to data from other satellites.[18] The main applications of the SPOT remote sensing techniques are, however, aimed at: (1) land use studies; (2) the assessment of renewable resources; and (4) cartographic work at medium scales, such as 1/100,000, the development of new types of maps, and the frequent updating of maps at scales of 1/50,000.[19]

Legal Aspects of Remote Sensing

The United Nations Committee on the Peaceful Uses of Outer Space and legal subcommittees have been the forum for discussions, debates, and negotiations between states on the legal problems relating to remote sensing. While attention was called to the use of remote sensing techniques as a means suitable for the planning of global resources by the Scientific and Technical Subcommittee as early as 1969, and the interdisciplinary Working Group on Remote Sensing of Earth by Satellites was established in 1971 as an organ within the subcommittee, it was not until 1974 that the subject of the legal implications of remote sensing from space was planned on the agenda of the Legal Subcommittee.[20]

With the aim of formulating draft principles relating to remote sensing the Legal Subcommittee began working on proposals for draft principles submitted by Argentina (1970), France (1969), the USSR (1973), Brazil (1974), France/USSR (1974), Argentina/Brazil (1974), and the United States (1975), together with input from other members, and by 1975 the Working Group was able to identify five common principles.[21] The number of principles now stands at seventeen.[22] On many of these principles, particularly the important ones, still no consensus has been reached, and the majority are still subject to alternative formulations.[23]

In any case, let us, without going into further details as to the historical evolution of the principles and states' attitudes toward them, proceed to examine the positions as reported by the Legal Subcommittee on the work of its twenty-first session (1–9 February 1982).[24] The divergent views in the discussion center upon the following main points: (1) the right to sense; and (2) the right to disseminate gathered data and analyzed information. These rights and how they should be exercised remain pivotal space policy issues.

The Right to Sense

Is any state free to conduct remote sensing activities by satellites and gather data on the natural resources of any other state? In other words, is the prior consent of the sensed state necessary for the sensing state to initiate remote sensing activities that will affect the territory of the sensed state? Positions taken by states range from complete prohibition of sensing without prior consent of the sensed state to the unrestricted freedom to conduct such activities. In between these positions there are states that agree with the right to sense without prior consent in principle, but put some qualifications on it.

In 1974, a joint proposal submitted by both Argentina and Brazil and supported by several Latin American and other developing countries called for the prohibition of initiating remote sensing activities that affect the territory of another state without the latter's prior consent.[25] Any unsanctioned sensing would give the sensed state the right to take all necessary measures authorized by international law.[26] The states advocating prior consent as a precondition for the sensing of their territory based their position on territorial sovereignty, including jurisdiction over natural resources and wealth, security implication, and economic considerations.[27]

The United States has been the main advocate of complete freedom to sense. It was supported in its position by the more advanced and industrialized countries like the United Kingdom and Japan. The United States based its position on two basic arguments, legal and technical. First, while conceding that remote sensing by satellites is partially an Earth-oriented activity, the United States considers remote sensing to be primarily a space activity and, therefore, subject to their 1967 Space Treaty, one of the treaty's fundamental principles being that outer space is free for exploration and use by all states.[28]

The United States has forcefully and repeatedly emphasized this point at relevant U.N. committee meetings.[29] Its position is supported by a 1973 United Nations Secretarial Background Paper, "Legal Implications of Remote Sensing of the Earth by Satellites," which came to the conclusion that (1) there does not appear to be any principle or rule of international law that makes it unlawful for a country to observe freely everything and anything in another country so long as it carries out its observations from beyond the limits of national sovereignty; and (2) the only restrictions are those contained in the obligation to act in accordance with the international law and to respect the corresponding interest of other states, as well as the duty to inform the United Nations secretary general and the public, to the greatest extent feasible and practicable, of the nature, conduct, locations, and results of national space activities.[30]

The technical argument of the United States is based on the fact that satellites are not able, despite their scientific sophistication, to detect invisible political boundaries though such ability would be necessary if the right to sense was premised on the prior consent of the many states scanned by a satellite.[31]

Other states, led mainly by the Soviet Union and France, admit the right of free sensing by satellites in principle but try to condition this right by some qualifications, though not in a uniform manner, to safeguard their national interest. Apart from prior notification, most of these qualifications relate to the restrictions to be imposed on the dissemination of gathered data and analyzed information, which will be discussed in the following paragraphs.

This position seems to have been the prevailing one in the discussion which took place in early 1982 at a meeting of the United Nations Legal Subcommittee.[32] Thus, at the very least, consensus has been reached on the right of free sensing in principle, but divergent views still remain as to whether this right should be subjected to any restrictions and if so what kind of restrictions should be imposed and how.

The Right to Disseminate Gathered Data and/or Analyzed Information

First of all, it should be mentioned that there has been agreement on the distinction between "primary data" and "analyzed information." "Primary data" means "those primary data which are acquired by satellite-borne remote sensors and transmitted from a satellite either by telemetry in the form of electromagnetic signals or physically in any form such as photographic film or magnetic tape, as well as preprocessed products derived from this data which may be used for later analysis."[33] "Analyzed information," though still to be clarified and interpreted by users, means "the end-product resulting from the analytical process performed on the primary data as defined . . . above combined with data and/or knowledge obtained from sources other than satellite-borne remote sensors."[34] Some states, led by the United States, have called for nonrestricted freedom to disseminate both primary data and analyzed information to anyone, including any third party. The United States stressed that any restrictions to be imposed on the dissemination of primary data and/ or analyzed information would render remote sensing by satellites technically and economically unfeasible. This argument was best expressed by United States Ambassador W. Tapley Bennet before the First Committee of the United Nations General Assembly (Political and Security) on 13 October 1975 when he stated:

> If the United States and other countries with such remote sensing satellites were to agree not to make available to third countries data on a sensed country without the latter's consent, we would in fact be able to share very little with anyone outside the United States, although it would be our intention to continue to make data available here. The natural swath of the satellite sensors commonly cuts across many national boundaries. The exercise of separating the billions of data bits along the lines of political boundaries is both financially prohibitive and scientifically disadvantageous. About such separation, in many parts of the world the consent of every country in a region might have to be obtained, through a time-consuming and complicated process which would ensure at the very least that the data released to countries without satellites would be much delayed and probably that it would be prohibited completely. There would be little incentive to pursue such a process.
>
> How, for example, could we or any other country continue to permit most other states to operate ground receiving stations under such a restrictive data-dissemination system? Normal coverage by a ground station is a circle approximately 3,000 kilometers in radius. For example, a station in the middle of South America could pick up data of at least part of every country on that continent. In other areas of the world it would be more; in some areas fewer. Under a restrictive data-dissemination proposal, we could not permit such a ground station to read out the data without the prior consent of all the countries in the region because the operator of that ground station would be a third country; that is, neither the sensed nor the sensing country.[35]

The United States believes that the free availability of facts advances world security and the better use of world resources; the Presidential Directive of 20 June 1978 stressed that "data and results from the civil space programs will be provided the widest practical dissemination to improve the condition of human beings on earth and to provide improved space services for the

United States and other nations of the world.''[36] This position is consistent with the United States Freedom of Information Act, which permits United States citizens to have access to sensed data.[37]

The stated position of the Soviet Union on this question is that

> every state is recognized to have the right to declare that certain types of primary data and analyzed information obtained by remote sensing of the earth with respect to its territory may be published or given to third States or natural or juridical persons of third states only with the express consent of the State making such a declaration. The declaration may relate to primary remote-sensing data with a spatial resolution of 50 meters or finer and to analyzed remote-sensing information obtained on the basis of such data. The dissemination of primary data and analyzed information obtained by remote sensing of the earth with respect to the territory of a State making such a declaration may be carried out only if the conditions stated in the declaration are observed.
>
> The declaration . . . shall be transmitted to the Secretary-General of the United Nations, who shall publish it for general information.[38]

Some states supported this position,[39] but others pointed out that the criterion of spatial resolution would not be feasible in remote sensing activities in view of technical and practical difficulties.[40]

There are, as previously mentioned, states that still hold the view that it is necessary that the dissemination of data and analyzed information be subject to the prior approval of the sensed states.[41] These states base their argument on the claim that dissemination without such prior approval would be contrary to the sovereignty of sensed states.[42] Colombia and Mexico, for example, are still among those states favoring a restrictive arrangement for the dissemination of primary data and analyzed information.[43]

Other states, while conceding the right of free dissemination of primary data and analyzed information, consider that in certain cases dissemination might be detrimental to the interest of states. Thus, they have sought such safeguards as holding any state conducting remote sensing activities internationally responsible for the dissemination of any primary data or analyzed information that might adversely affect the national interest of a sensed state.[44]

At issue is the sovereignty of states over their natural resources and its limitations. The question of whether sovereignty over natural resources extends to data and information about those resources is still a debated subject.[45] What might further complicate the matter is the fact that primary data and analyzed information are substantially different. While it can be argued that primary data collection might violate a state's sovereignty over its natural resources, it is also necessary for the sensing state to process the primary data so as to convert it into useful information, which means that the sensing state would have substantial input into the analyzed information.[46] It would in this way gain even more control over knowledge of the data pertaining to the sensed country.

Future Prospects

There is hope for international cooperation in the field of remote sensing. Many draft principles point in this direction, particularly in relation to disaster prevention and relief operations as well as the protection of the natural environment of the Earth.[47] Other principles may also contribute to cooperation, such as those that call on sensing states to give other states opportunities for participation in remote sensing programs and technical assistance.[48] Another hope is that developing and poorer countries will become more aware of remote sensing techniques and their own potential benefits. United Nations training seminars play a very useful role in this context.[49] The most practical steps are programs like the United States Landsat and the French project SPOT. Such programs, it is hoped, will overcome all political, technical, and economic problems that may constitute obstacles to international cooperation in this field.

As for the legal issues, it is acknowledged that space activities have challenged some basic principles of international law like sovereignty and its limitations and implications, but this is a price we have to pay for more advanced technology. After all, it is clear that throughout history advances in science and technology bring challenges to established values. Often the result is not only the adaption of old values to new situations, but also the abolition of old values and the emergence of new ones.

The point to be made is that we should become more realistic and concentrate on cooperation instead of wasting our time trying to protect some of our expedient national interest disguised under various legal principles and concepts. There is no other alternative. Our world has become a very small ship and we have to put up with one another and learn to live together or the ship might sink.

Notes

1. See, e.g., U.N. Doc. A/AC.105/280 (22 December 1980), p. 7.
2. See, e.g., ibid., pp. 6–13.
3. Ibid., pp. 5–6.
4. See, e.g., U.N. Doc. A/AC.105/296 (18 November 1981), pp. 5–6.
5. See, e.g., U.N. Doc. A/AC.105/297 (30 December 1981), pp. 26, 27, 33, 34, 36.
6. For a complete list of remote sensing activities concerning this regard in Bangladesh, Indonesia, Pakistan, and United States, see ibid., pp. 2–37.
7. See, e.g., Jean-Louis Magdelénat, "The Major Issues in the 'Agreed' Principles on Remote Sensing," *Journal of Space Law* 9, nos. 1, 2: 112.

8. Ibid.
9. U.N. Doc. A/AC.105/292 (17 July 1981), p. 4.
10. *Aviation Week and Space Technology*, 14 June 1982, p. 90.
11. For more details, see ibid., p. 87.
12. For details, see statement of Dr. Anthony J. Calio of NASA before the Subcommittee on Space Science and Applications of the House Committee on Science and Technology, 96th Cong., 1st sess., 16–18 October 1979 (Washington D.C.: Government Printing Office, 1979) vol. 1, pp. 14–19; U.N. Doc. A/AC.105/ 297, pp. 2–37.
13. Statement of Dr. Anthony J. Calio, ibid., 15; SPOT *Newsletter*, 14 December 1981, p. 5.
14. *Aviation Week and Space Technology*, 11 January 1982, 98; *France Espace*, CNES International News Letter (April 1982): 4.
15. *France Espace*.
16. SPOT *Newsletter*, 14 December 1981, p. 5.
17. Ibid.
18. Ibid.
19. U.N. Doc. A/AC.105/292, p. 5.
20. Magdelénat, "Major Issues in the 'Agreed' Principles on Remote Sensing," p. 113.
21. Ibid.
22. Ibid.
23. U.N. Doc. A/AC.105/305, 24 February 1982, Annex 1, pp. 7–11.
24. Ibid., pp. 1–21
25. Article V of the draft proposal, U.N. Doc. A/AC.105/C.1, 1047 (1974)
26. Article VI of the draft proposal, ibid.
27. See, e.g., Magdelénat, "Major Issues in the 'Agreed' Principles on Remote Sensing," p. 114.
28. Ibid., p. 115.
29. Ivan A. Vlasic, "Principles Relating to Remote Sensing of the Earth from Space," in *Manual on Space Law*, vol. 1, ed. N. Jasentuliyana and R.S.K. Lee, (Dobbs Ferry, N.Y.: Oceania Publications, 1979), pp. 303–22.
30. U.N. Doc. A/AC.105/118 (1973).
31. Magdelénat, "Major Issues in the 'Agreed' Principles on Remote Sensing," p. 115.
32. U.N. Doc. A/AC.105/305.
33. Ibid., Annex 1, p. 7, Principle 1.
34. Ibid.
35. Quoted in Vlasic, "Principles Relating to Remote Sensing of the Earth from Space," pp. 322–23.
36. Quoted in Magdelénat, "Major Issues in the 'Agreed' Principles on Remote Sensing," p. 118.
37. Ibid.
38. U.N. Doc. 105/305, Annex 1, p. 18.
39. Ibid., p. 5.
40. Ibid., p. 6.
41. Ibid., p. 5.
42. Ibid.
43. Ibid., pp. 12, 15.
44. Ibid., p. 5.

45. Ibid., pp. 5–6.
46. Magdelénat, "Major Issues in the 'Agreed' Principles on Remote Sensing," p. 117.
47. Principles V and VIII, U.N. Doc. A/AC.105/305, Annex 1, pp. 8–9.
48. Ibid., Principles IV, VI, and X.
49. See, e.g. U.N. Docs. A/AC.105/280, A/AC.105/295, A/AC.105/296, and A/AC.105/300.

11

The Moon Treaty: Reflections on the Proposed Moon Treaty, Space Law, and the Future

Nathan C. Goldman

Ever since that early primate picked up a sturdy branch to defend itself, humanity has struggled to conquer nature—and to govern it, often the more difficult of the two tasks. As we spread out over the globe exerting our will, we molded societies, governments, and laws to benefit from and organize our conquests. Now that humanity is beginning to move into space, to conquer at least our little solar system, we necessarily bring our social and legal order with us. A mature legal order for the solar system will signal the human conquest of the planets.

With the 1957 launch of Sputnik, humanity grasped the sturdy branch to conquer space. Within a year, the world made the first moves to govern the new conquest. The United Nations quickly established the Committee on the Peaceful Uses of Outer Space (hereafter COPUOS) with legal and technical subcommittees.[1] Initially composed of 18 nations, the committee's growth to 53 nations reflects the increasing concern with and use of space by the world community. To enhance the legitimacy of any new legal order, the United Nations authorized COPUOS to report out only those treaties obtaining unanimous consent. One vocalized objection spoils the consensus.[2] Within this framework, the Western, Communist, and Third World nations have negotiated the beginnings of a legal order to regulate the activities of humankind beyond the earth. In so doing, conflicts among these blocs have been adjusted but not solved.

Conflict notwithstanding, the U.N. committee and the General Assembly have passed five compromise space treaties, composing a skeletal international law of outer space. The United States has ratified four of these treaties. The fifth, the proposed Moon Treaty, has been signed or ratified by a dozen nations. This proposed Moon Treaty provided that the treaty would become binding on its parties when five nations signed and ratified it. By December 1983, four nations had ratified the treaty: Chile, the Netherlands, the Philippines and Uruguay. Signatories that have not ratified the treaty include France, Austria, Rumania, Peru, Morocco, India, and Guatemala.

In 1970, Argentina had proposed the first draft of this treaty to the Legal Subcommittee of COPUOS. Other nations, including the Soviet Union in 1971, proposed other drafts of a treaty dealing with the moon. Ironically in the context of the later U.S. Senate debate on the treaty, the Soviet Union at that time wanted the document to apply only to the moon. Likewise, the Soviets opposed the "common heritage" language and believed that provisions dealing with space resources were premature.

After years of impasse, the COPUOS reached consensus on the compromise draft treaty governing activities on the moon and other celestial bodies; the date was 3 July 1979. The U.N. General Assembly approved the treaty later that same year.[3]

The U.S. Space Lobby and the Defeat of the Moon Treaty

Into this delicately balanced international arena the U.S. space lobby emerged to battle the proposed Moon Treaty. Among the 50 to 100 U.S. prospace groups, the L-5 Society is small but one of the most active. (L-5 refers to one of the L- or Langrangian-points. La Grange was a French mathematician who developed the equations for the balancing of the gravities of three objects: the sun, moon, and Earth. One of the equations identifies a point of constant equilibrium. Anything, including a space colony, put there stays there.) The society, with 10,000 members, is dedicated to a permanent colony in L-5 and to space industrialization in general. Its single-minded dedication to the human colonization of space is the source of its cohesion and drive.[4] In late 1979, its Board of Directors decided, over some opposition, to actively oppose ratification in the U.S. Senate of the proposed Moon Treaty.

To this end, the L-5 Society hired attorney-lobbyist Leigh Ratiner, of Dickstein, Shapiro and Morin. Ratiner, who had worked for Kennecott Copper Corporation and opposed the partially analogous Law of the Sea negotiations, was well prepared to lead the opposition to the proposed Moon Treaty. He and Mark Hopkins, an economist at Rand Corporation and a leader in L-5, orchestrated the political battle. Soon, L-5 had enlisted the American Astronautical Society, *Future Life*, and *Omni* in opposition to the treaty. Stan Rosen, chairman of the Public Policy Section of the Los Angeles branch of the American Institute of Aeronautics and Astronautics, was quoted in the newpapers: "We're afraid it [the treaty] would act as a disincentive to private investment in the exploration or the development or the exploitation of lunar materials or any resources which might occur in the solar system."

The battle turned in L-5's favor when the space industry began to express opposition to the treaty. Julian Levine spoke for the Aerospace Industries Association, the central organization of the space enterprises: "We understand that there needs to be a legal format for the exploration and extraction of items from space. We just don't think the specifics of this treaty protect America's self-interest."

United Technologies, Inc., joined the antitreaty movement with several advertisements that read:

> In its draft form, the treaty is inimical to America's interests. It would frustrate the access of our nation and our people to space for purposes of industrialization. If the U.S. becomes party to the agreement, there's scant prospect of private enterprise's ever being able to develop the resources of the moon and outer space to improve life on Earth.

The American Bar Association advocated that the treaty be ratified but only if the Senate passed a series of reservations. These reservations would be the binding U.S. interpretation of the treaties, and we would expect other treaty parties to accept the interpretation.[5]

In July 1980, Ratiner and Gerald Driggers, then L-5's president, testified that the proposed Moon Treaty in space and the Seabed Treaty in the oceans would endanger private enterprise. Senators Frank Church (D-ID) and Jacob Javits (R-NY), leaders of the Senate Foreign Relations Committee, warned then-Secretary of State Cyrus Vance not to submit the treaty to the Senate. Soon after, the Senate and the White House tabled the treaty indefinitely.[6]

The victory was wholly L-5's. The society publicized the issue and mobilized an alliance of space and conservative groups. In 1980, a Democratic Senate refused to vote on the treaty. The Republican-controlled Senate since 1981 has been no more considerate. L-5 not only succeeded in its lobbying effort but received recognition in Washington.

Other segments of the legal and space communities, however, questioned the methods, if not the goals, of the organization. To them L-5 had lost much credibility on Capitol Hill. Rather questionable interpretations of international law and conventions left some with the opinion that the society has responded with its heart more than its head. This is unfortunate because the treaty as well as the policy process may need revisions and because the space community has a major role to play in the formulation of U.S. space policy and in the creation of an international regime that will allow free enterprise to develop without the risks of international chaos.

The senatorial debate that reargued the issues of the COPUOS debate highlighted the lack of participation by private industry and groups in the diplomatic negotiations at the United Nations. The conflict in COPUOS and the Senate centered on three provisions: (1) the common heritage of mankind; (2) the international regime and exploitation of space resources; and (3) whether the treaty created an interim moratorium on the mining of space resources.

The Common Heritage and the International Regime

The concept of the common heritage lies at the heart of future economic and political relations between the great powers and between the developed and developing worlds. It originated in a clause of the 1967 Outer Space Treaty that reserves space as the "province of all mankind," "for the benefit of mankind." In the 1970s the negotiations on the Law of the Sea adopted, rephrased, and embellished this concept. In the draft treaty on the seabeds, "common heritage" had grown to include technology transfer and other sharing among nations as well as an international regime to regulate exploitation of the ocean bottom. The U.S. position is that in space, however, "common heritage" refers only to equal access to space for all countries—not to the Third World's proposed International Economic Order.

Despite the rhetoric of some opponents to the treaty, the "common heritage of mankind" in the Moon Treaty is not the same as that term in the Law of the Sea Treaty; the "common heritage" here means that the states "undertake to establish an international regime . . . to govern the exploitation of natural resources of the moon"[7] in accordance with the common heritage principle.

The opposition to this portion of the Moon Treaty has stressed two points: (a) the analogy to the "common heritage" of the Law of the Sea Treaty; and (b) Third World-Soviet assault on private enterprise. Neither point is well taken. To break its impasse of a decade, COPUOS accepted the Soviets' caveat that the "common heritage" would be interpreted only in the context of *this* treaty.[8] Although the Soviets inserted this provision to prevent reference to human rights interpretations of "common heritage," this caveat also excludes reliance on the Law of the Sea Treaty's common heritage history.

This limitation on the "common heritage" becomes even more important for interpreting the international regime which the parties must "undertake to establish under this Treaty."[9] Article XI, which obligates the parties to try[10] to establish an international regime, documents the right of countries (and their nationals) to receive a profit for their "efforts":

> (d) An equitable sharing by all States Parties in the benefits derived from those resources, whereby the interests and needs of the developing countries as well

as the *efforts of those countries which have contributed either directly or in-directly to the exploration of the moon shall be given special consideration.*
[Emphasis added.][11]

"Equitable" is not equal, and allows for a profit.

A related problem associated with an international regime, controlled by the Third World, is whether the private company will be allowed enough profit to make investment in the new space ventures worthwhile. In the establishment of any regime, the U.S. negotiators will have a major role in defining the regime's powers and limitations. If the U.N. committee continues to operate by consensus, the United States can veto any formula that might jeopardize U.S. interests in the new regime. Even if fees or royalties on new space ventures are substantial, U.S. tax laws can be revised to assist these private efforts. Already U.S. companies, such as those in the oil industry, deduct foreign taxes from their domestic income taxes. Fees for space exploitation could similarly be deducted.

Finally, opponents of the treaty contend that the international space regime would follow the model proposed in the draft Law of the Sea Treaty. That model would create a "General Assembly-type body dominated by the Third World."[12] This argument ignores possible alternative regimes.

INTELSAT is at least one alternative paradigm that allows private profit and incentive. The United States owns 24 percent of INTELSAT and has 24 percent of the vote on many issues. Comsat Corporation represents the United States and until recently managed the consortium.[13] The INTELSAT arrangement does not require equal sharing, nor does it prohibit private activity. Inmarsat and even ESA (the European Space Agency) are other "equitable" arrangements.

"In Place"

Opponents argue that the proposed treaty prohibits any ownership of space ores or resources,[14] citing Article XI, Section 3, for this proposition:

Neither the surface nor the subsurface of the moon, nor any part thereof or natural resources in place, shall become the property of any state, international, inter-governmental or non-governmental organization, national organization or non-governmental entity or of any natural person.[15]

During the U.N. committee debates, however, the U.S. representatives inserted an official interpretation of these words:

The words "in place" . . . are intended to indicate that the prohibition against assertion of property rights would not apply to national resources once reduced to possession through exploitation either in the pre-regime period or, subject to the rules and procedures that a regime would constitute, following establishment of the regime.[16]

"These statements by the United States drew no response, and this silence is once again a part of the history of the treaty."[17] The silence is extremely important to the status of this clause and its interpretation in law. Testifying before a Senate subcommittee, Robert Owens of the U.S. Department of State reasserted the American position:

> Although there can't be assertions of ownership of the resources of the moon before extracted from the moon, ownership can be asserted at that point. . . . As a legal matter, the preparatory work of a treaty and the circumstance of its conclusions are, of course, recognized by the Vienna Convention on the Law of Treaties to be a supplementary means of interpretation to be resorted to where the meaning of provisions is ambiguous or obscure.[18]

More important, Owens went on to say:

> As a matter of customary treaty law, if other State Parties to a Treaty do not contest such declaration [reservations] within a reasonable time, the declarations become an integral part of the treaty relationship between the State making the declaration and each non-objecting state.[19]

Later in the subcommittee hearing, Ronald F. Stowe, chairman of the Aerospace Law Committee of the Section of International Law of the American Bar Association, made a similar argument. The nations that object to our reservations simply do not have a treaty with the United States. Thus, this objection to the treaty is without merit if the Senate ratifies the treaty and restates the reservations.[20]

Moratorium

Many opponents of the Moon Treaty emphasize that commercial exploitation of space is subject to a de jure moratorium pending the establishment of the international regime. Once again, the position of the United States during the negotiations, which no party contradicted, was that no moratorium would exist.[21]

On the legal level, the United States admits no moratorium. On the practical level, moreover, the U.S. government could provide legal and economic support to prevent any de facto moratorium. If the international regime were hostile to private interests, guaranteed government insurance or reimbursement could blunt the hostility. Indeed, in preparation for the Law of the Sea Treaty, Congress considered this logic in the Deep Seabed Hard Mineral Resources Act, P.L. 96–283. And to repeat, the fees or royalties assessed by the international regime could be treated as an offset against domestic income taxes.

Finally, some opponents of the treaty have argued that the treaty's ambiguity creates a de facto moratorium on the exploitation of space resources because entrepreneurs cannot know their real economic risks. It should be pointed out that without this proposed treaty, the United States is still a party to the 1967 Outer Space Treaty, which is at least as ambiguous and has the same "non-appropriation" of space clauses.[22] One opponent declared, "Today the United States has a perfect right to exploit space resources for profit. . . . It should be stressed that the Moon Treaty does not create even a single new right beyond those the United States already enjoys under existing international law."[23] The 1967 treaty states, however: "Outer Space, including the moon and other celestial bodies, is not subject to national appropriation by claim of sovereignty, by means of use or occupation or by any other means."[24] If one ignores the official U.S. interpretations of the meanings of this article, the "clear meaning" already prohibits mining the celestial spheres.

Likewise, Article I of the 1967 treaty provides that space shall be used "for the benefit and in the interest of all countries, . . . and shall be the province of all mankind."[25] Is this the common heritage of mankind, a rose by any other name? Compared to the 1967 treaty (especially as shorn of U.S. interpretations) this right in the Moon Treaty to use space resources for science and to build space bases and commercial mining prototypes are positive grants of rights.[26]

Problems

Although we find much opposition to the Moon Treaty to be misdirected, the treaty does present at least two other obstacles to human use of space. It has an environmental provision that blocks the eventual terraforming of a planet, that is, making Earth-like or inhabitable the surface of a planet. More important, because the treaty applies to all celestial bodies, including asteroids, the provision could jeopardize the mining of asteroids. It reads:

> In exploring and using the moon, State Parties shall take measures to prevent the disruption of the existing balance of its environment whether by introducing adverse changes in such environment, its harmful contamination through the introduction of extraenvironmental matter or otherwise.[27]

When an asteroid has been transported to Earth's orbit and half of its mass is consumed as ore or debris, can it be said that the asteroid's environment has been "disrupted"? Likewise, in mining the surface of a planet, at what point do the waste products in the atmosphere and the surface scarring equate the "disruption of the existing balance of its environment"?

Probably less important is the problem with a provision that a "State Party establishing a station shall use only that area which is reasonable for the needs of the station."[28] In the competition for useful sites on celestial bodies, the first nation to establish a base would probably have an expansive concept of its "reasonable" needs. This provision only beclouds an already hazy question.

Conclusions

The basic criticism against the treaty remains: it is vague and may be more impediment than implement to the rational development of space. Because the present treaty is unlikely to be ratified either by the United States or by the USSR, one alternative may be to ignore it (even if ratified by five nations). Then, its positive provisions (use of space resources for bases, experimentation, and commercial prototypes) could be amended to the 1967 Space Treaty. Parsing the more difficult international regime and resource allocation provisions can await a time nearer the actual event when the technology and politics are more certain.

In the meantime, the Moon Treaty and its aftermath leave several important questions:

1. Is the consensus procedure used by COPUOS still viable? When the committee was established in the late 1950s, it had only 18 members (now it has 53); then the issues were less pressing to fewer countries. One wonders whether the requirement of unanimity among so large a group is a direct cause of the treaty's ambiguity.
2. Is the lack of a coherent U.S. space policy also responsible for the ambiguity of the treaty? It seems that the national failure of vision regarding the exploitation of space has been detrimental to U.S. negotiations in committee proceedings. This lack of foresight can be seen in the whole of U.S. space planning.
3. What role, then, should the space community play in developing space policy and in negotiating future arrangements on space law? One of the most disconcerting facts brought out during the treaty hearings was the lack of private-sector participation in negotiating the treaty.[29] (This lack should be juxtaposed against the extensive role of the private sector in the Law of the Sea Treaty negotiations.)

If the United States continues to lack the foresight or the fortitude to defend its interests in space, then the opponents of the treaty may be right after all: the United States may not stand up for its own interpretations of the treaty. Of course, if U.S. diplomacy were to take that weak stand, U.S. industry would never take the risk to go into space in the first place.

With the growth of the citizen space lobby and the coming age of private space business (especially in satellites),[30] the space community is directly affected and has much to offer to the debate on national and international space policy. Space advocates must conduct a genuine, basic debate on the goals and methods for exploring outer space. I submit, however, that the basis for any stable, successful exploitation must be rooted in law. In thirty years, when the United States is still likely to be the only nation capable of mining the moon, a dozen other space-faring countries will be physically able to interfere with the effort. Without international order, the risks to peace are great. The risks for free enterprise are even greater; without legal and political security, private capital will not go into outer space. The question, therefore, is not the fist of regulation versus the invisible hand. It is whether such regulation will enable capitalist development in space.

Earlier space treaties, such as the Liability Convention, do ignore or inhibit the private development of space. The convention makes states responsible for the actions of their nationals (corporations) and imposes an absolute liability standard on damages on Earth caused by space activities. This provision was unnecessary because space will continue to be an ultrahazardous activity subject to absolute tort liability for a long time. The treaty, unfortunately, prevents the natural evolution of space exploration into a normal fault-liability status. It should be amended.

After blocking the Moon Treaty, some space advocates are seeking U.S. withdrawal from this and other space conventions. That regrettable threat should only be our last resort. The U.S. government and its private citizens must fight this battle in the international arena of law. We must negotiate for an international order that will permit our system to compete fairly in space against the others. We cannot hide behind an isolationist-rejectionist attitude at home and also aim for the stars.

Notes

1. P.C. Jessup and H.J. Taubenfeld, "U.N. *Ad Hoc* Committee on the Peaceful Uses of Outer Space," *American Journal of International Law* 53 (October 1959): 77.
2. E. Galloway, "Consensus Decision-Making on the U.S. Committee on the Peaceful Uses of Outer Space," *Journal of Space Law* 7 (Spring 1979): 3.
3. U.S. Senate, Committee on Commerce, Science and Transportation, *Space Law: Selected Basic Documents*, 2d ed., (1978); Treaty on Principles Governing the Activities of States in the Explorations and Use of Outer Space (hereafter cited as *Treaty on Principles*), 10 October 1967; Agreement on the Rescue of Astronauts, the Return of Astronauts and the Return of Objects Launched into Outer Space, 3 December 1968; Convention on International Liability for Damage Caused by Space Objects, 9 October 1973; Convention on the Registration of Objects Launched into Outer Space, 15 September 1976; see The Foundation,

Commercial Space Report, Gary Hudson; Nicolas M. Matte, "The Draft Treaty on Moon: Eight Years Later," *Annals of Air and Space Law* 3 (December 1978).

4. The *L-5 Newsletter*, published in Tucson, Arizona, since 1975 has recorded the growth of not only the society but also the entire citizen space movement. Other publications that document this phenomenon include *Space Age Review* (California) and *Insight* (National Space Institute, Washington, D.C.). The two other important space groups are Carl Sagan's Planetary Society (120,000 members) and the National Space Institute, originally sponsored by the late Wernher von Braun. The other space groups range from professional aerospace engineering organizations and political action campaigns (PACs) to science-fiction and near-cult groups.

5. Carl Q. Christol, "The American Bar Association and the 1979 Moon Treaty: The Search for a Position," *Journal of Space Law* (September 1981): 77; *Times-Union* (Albany), 27 April 1980; *Los Angeles Times*, 18 April 1980.

6. *The Moon Treaty: Hearings Before the Subcommittee on Science, Technology and Space of the Committee on Commerce, Science and Transportation*, U.S. Senate, 96th Cong., 2d session., 29–31 July 1980; *Los Angeles Times*, 18 April 1980; *L-5 Newsletter*, July–September 1980; *L-5 Newsletter*, 7 July 1980.

7. "An Agreement Governing the Activities of States on the Moon and Other Celestial Bodies" (hereafter cited as *U.N. Moon Treaty*), U.N. Document A/AC 105-L 113/add. 4, 1979, Article XI, paragraphs 1,5.

8. Art Dula, "Free Enterprise and the Proposed Moon Treaty," *Houston Journal of International Law* 2 (1979): 3, 9.

9. *U.N. Moon Treaty*, Article XI, paragraph 5.

10. "Undertake" also connotes that the effort may fail.

11. *U.N. Moon Treaty*, Cert. XI, paragraph 7(d).

12. Bennet, James C., "The Second Space Race," *Reason* 12 (1981): 21–31; Dula, "Free Enterprise," p. 17.

13. See International Telecommunications Satellite Organization Agreement with Annexes, Operating Agreement Relating to the International Telecommunications Satellite Organizations, Communications Satellite Act of 1962, as amended, in *Space Law: Selected Basic Documents*; testimony of Robert Owens et al., *The Moon Treaty*.

14. See testimony of Gerald Driggers and Leigh Ratiner, *The Moon Treaty*; also Dula, "Free Enterprise," p. 11.

15. *U.N. Moon Treaty*, Article XI, paragraph 3.

16. Dula, "Private Enterprise," pp. 11–12, citing the statement of Neil Hosenball, NASA chief counsel.

17. Ibid., p. 12.

18. Testimony of Robert Owens, *The Moon Treaty*, pp. 6–7, 10.

19. Ibid., pp. 16–17.

20. Testimony of Ronald F. Stowe, ibid., p. 81; the ABA suggested that the Senate pass this treaty with such reservations; see Christol, "The American Bar Association and the 1979 Moon Treaty," pp. 77–92.

21. Dula, "Private Enterprise," p. 10.

22. *Treaty on Principles*, Article II.

23. Dula downgrades the status of such interpretations in his article.

24. *Treaty on Principles*, Article II.

25. Ibid., Article I.

26. *U.N. Moon Treaty*, Article VI, paragraph 2.

27. Ibid., Article VIII, paragraph 1.
28. Ibid., Article XI, paragraph 1.
29. Testimony of Dr. Charles Sheffield, *The Moon Treaty*, pp. 85–88; testimony of Neil Hosenball, ibid., pp. 52–53.
30. Nathan Goldman and Michael Fulda, ''The Outer Space Lobby and the 1980 Elections,'' paper presented at the Southern Political Science Convention, November 1981; Nathan Goldman and William A. Good, ''Commercial War in Space: The Battle for Markets,'' unpublished manuscript, 1981.

Part III

The Sociology of Outer Space

12

Beyond Bureaucratic Policy:
The Space Flight Movement

William Sims Bainbridge

The aim of the space flight movement—exploration and colonization of the universe—is so vast and revolutionary that it cannot be achieved by the ordinary operation of day-to-day social forces and institutions. Consequently, we must be prepared to think in imaginative ways if we are to understand how this ''giant step'' in human history may come about.

One critical fact to consider when examining the prospects for space colonization is that we are not in contact with an extraterrestrial civilization. Even if only one star in a million produces intelligent life, a hundred thousand civilizations would appear in our galaxy. Simple calculations suggest that a single colonizing species could fill the galaxy in a few million years, far less than the time since civilization probably began to appear.[1] Thus, either these calculations are wrong or interstellar colonization must be exceedingly difficult and unlikely.[2] Yet technology capable of achieving it can already be sketched. Therefore, the apparent lack of colonization may be due to social rather than technological considerations.[3]

As technology advances, we can better fulfill our needs by reengineering not only our environment but also our species itself. Birth control seems to solve population pressures better than does interplanetary colonization.[4] A static industrial base might be more satisfactory than the heroic mining of the asteroids. Direct intervention in human genetics and brain physiology can end crime and deviance, even to the point of reducing mankind to automatons.[5] Bread can be manufactured from sewage, and circuses can be simulated with computer graphics. Once a species has the power to transform itself to satisfy itself, it has no utilitarian need to explore and conquer the universe. The threat of self-annihilation, whether through nuclear war or some other misapplication of science, may be overcome. A species might establish an absolute cultural and political "freeze" in the service of its own preservation. Perhaps only highly unlikely and risky accidents, occurring just before a species arrives at such a stasis, can propel it out across the gulf of space to the stars.

The Space Flight Movement

In the issue of *Time* magazine marking the tenth anniversary of the first moon landing, science fiction writer Arthur C. Clarke gave additional support for a main finding of my 1976 study of the space flight movement: "Space travel is a technological mutation that should not really have arrived until the 21st century. But thanks to the ambition and genius of Wernher von Braun and Sergei Korolev, and their influence upon individuals as disparate as Kennedy and Khrushchev, the moon—like the South Pole—was reached half a century ahead of time."[6] A similar view was expressed by Bruce Murray, until recently head of NASA's Jet Propulsion Laboratory: "I think Apollo was an anomaly—we were lucky. It happened four decades too early, and Voyager and Viking and the rest were all spinoffs from that."[7] These leaders of the space flight movement express the faith that space travel is a natural development for humanity which was speeded up by the activism of their movement. But my research on the social history of rocketry suggests to me that the moon never would have been visited at all were it not for the perilous machinations of the movement. Interplanetary colonization and interstellar exploration demand similarly astonishing miracles.

But the space flight movement is now fractionated and not accomplishing very much, quite in contrast to the great milestones it achieved in the 1960s. The current malaise stems partly from the movement's success. What began as the *parallel behavior* of isolated intellectuals evolved through a phase of *collective behavior* as a minor fad and organized itself into a true social movement. But now it has been greatly absorbed into societal institutions.[8] To the extent that it has been captured by conventional bureaucracies, the space flight movement can no longer promote revolution.

It began almost simultaneously in four great nations: Germany, the Soviet Union, Britain, and the United States. First, individual theorists like Tsiolkovsky, Goddard, and Oberth developed the principles of space travel. This was the phase of *parallel behavior*—individuals doing essentially the same thing but without any communication among them. Then a cascade of books, articles, and lectures established networks of communication and a shared space culture. This was the phase of *collective behavior*—individuals influencing one another but without formal planning and organization. Soon the moderate level of organization that defines a social movement was achieved in the founding of amateur space travel clubs in Germany (1927), the United States (1930), the Soviet Union (1931), and Great Britain (1933). In most respects this was an elite rather than a mass movement. The founders were an intellectual elite but generally without much personal power or wealth.

The German Case

The German club experimented with simple liquid fuel rockets and grew to include a thousand members.[9] But all attempts to find sufficient resources to build a real spaceship seemed doomed to failure. Corporations put in a little money, then lost interest. Where was the profit to be made in space? True, corporations are interested in near-Earth space today, but only because a start-up investment of $100 billion has already been made by government to develop the technology. And the general public will not even support modern charities adequately, a job largely taken over by government, let alone raise through voluntary contributions the immense sums required for space-flight. But as it died in the deepening economic collapse of the Great Depression, the club found a strategy that would work: *a military detour.*

During the Nazi era, a young aristocrat, Wernher von Braun, persuaded the German military to build the V-2 rocket; Apollo's Saturn V booster, which took a man to the moon, was its direct descendent. Again and again, von Braun and other leaders of the movement essentially succeeded with trickery as the tactic to get societal leaders to invest in space technology. The precondition for success was a powerful patron locked in combat with an opponent who had just achieved an advantage over the patron. The *spaceman,* a movement leader like von Braun, went to the patron offering his rocket technology as a way of outflanking the opponent to gain a counteradvantage. The patron had to be sufficiently ignorant of the proposal's technological difficulties, and hold sufficient independent power, that his pressing needs could drive him to sponsor the spaceman's project. Whether the new technology would really help the patron or not—and often it could not—it would further the cause of space.

Consider the German army between the wars. Defeated by the Allies and severely restricted by treaty, it could not defend its own nation from the French invasion of the Ruhr in 1923. The agreements ending World War I restricted German artillery but said nothing about rockets. Thus, the army allowed itself to be convinced by von Braun to develop long-range liquid fuel rockets as a way of redressing the great disadvantage it suffered with respect to the Allies. This was a conscious trick by von Braun. The right technology for most military purposes involves solid fuels, requiring solution of a very different set of technical problems. While von Braun's V-2 did kill more than 4,000 persons, I believe that these resources would have served the Nazis much better if they had been invested in the Me-262 jet interceptor, the crude

V-1 buzz bomb, or even in ordinary armaments. It was not until the 1950s, when Soviet and U.S. versions of the V-2 could carry nuclear warheads, that a cost effective weapon emerged. Today most U.S. missiles use solid fuels, which are poorly adapted for space purposes.

After the war, von Braun and others performed similar maneuvers to gain the support of Kennedy and Johnson, as did the Soviet branch of the movement in exploiting the fears and military weaknesses of Stalin and Khrushchev.[10] It is noteworthy that the man who led the Russian space program that developed the Sputnik launcher was Sergei Korolev, president of the entirely amateur Soviet space club in the early 1930s.

The space flight movement exploited military and political tensions, in far more cases than I could list here, to move toward its goal of interplanetary flight and colonization. This was a tightrope walk across a moral abyss. It would be nice if social and cultural progress resulted from clear, consensual decisions openly and freely made by all people of the world in an atmosphere of good will. But that is not how history works.

The Current Malaise

The successes that led to a lunar landing have not been matched in the post-Apollo period. Neil Armstrong garbled his great words about his "one small step . . ." as if the present confused situation of spaceflight demanded a confused prophecy. For the past fifteen years, unlike earlier years, there have not been good opportunities to exploit and it is hard to see how they will come again.

In great measure, the movement has become institutionalized, an adjunct to the military-industrial complex. There is no free amateur club left in the Soviet Union, and the U.S. club has evolved into the American Institute of Aeronautics and Astronautics (AIAA), the paramount aerospace technical society, two-thirds of whose members are from the government or corporations. But the movement lives, in a swarm of small, marginal organizations. In 1953, space enthusiasts dissatisfied with the increasing conservatism of the AIAA founded the American Astronautical Society. In 1981, Trudy E. Bell was able to tabulate 11 trade or professional space groups, 39 independent space-boosting groups, and a further 9 space-sympathetic groups. By 1984 this total was even larger. Bell intentionally excluded from consideration the less respectable mystical, religious, or countercultural wings of the movement, and did not count foreign space societies. Although the citizen groups mainly do inexpensive volunteer space-boosting, their aggregate annual budget had by 1980 already passed $4 million.

Some of these groups have developed elaborate, technically detailed plans for future space projects, notably the L-5 Society, which wants us to build floating cities at the vertex of an equilateral triangle based on the Earth and moon. Perhaps because of the vast cost of such an undertaking, L-5 has been unable to provide a reasonable economic prospectus or a workable political plan to bring its dreams to realization.

Far more realistic, but unlimited in its ultimate aims, is the Planetary Society, founded by scientists Carl Sagan and Bruce Murray, with now over a hundred thousand members. The introductory brochure says, ''The Planetary Society is devoted to encouraging, supporting and participating in the greatest adventure the human species may ever know—the exploration of the solar system, the search for planets around other stars and the quest for extraterrestrial life and intelligence.'' Through continual media events and by sponsoring modest technical projects, the society believes it can inch humankind along the long road to the society's goals.

Some groups even attempt to develop their own space technology. A few years ago, the Committee for the Future tried to buy a surplus Saturn V launch vehicle and plant a scale-model colony on the moon. Although the patrons of the committee were wealthy, the project fell through, and it has become apparent that no private group can currently garner enough resources for spaceflight. But perhaps smaller contributions can be made. The Independent Space Research Group is preparing an amateur space telescope for launch by the Space Shuttle. Ham radio enthusiasts have participated in the space program since the beginning, not only monitoring artificial satellites but even getting some modest ones of their own into orbit when larger payloads were being launched.

New Initiatives Promoted

In 1981, the World Space Foundation unveiled—or more accurately, unfolded—the world's first prototype solar-sail space vehicle. Partly financed by the Charles A. Lindbergh Fund, this project falls short of providing a functioning interplanetary probe, but it does demonstrate a direction engineers might go to develop a cheap means of propulsion.

Several of the more sober organizations collect money for astronomical research projects modest enough to be funded privately yet capable of deepening our understanding of the universe and providing the information necessary for an interplanetary civilization. For example, astronomers at the University of Arizona have a series of projects to find, catalogue, and study asteroids, the small rocky worlds that orbit in several groups outward from Earth in the solar system. They have sought support from the Planetary Society and the Space Studies Institute. The World Space Foundation also contributes to asteroid studies.

Money Problems and the Search for Life

Perhaps more exciting is the project at the University of Pittsburgh to renovate an existing telescope so that it can make the precise measurements of star positions required to deduce the existence of planets in other solar systems. This is difficult work, and prior reports of such discoveries, published over the past forty years, are currently discounted by astronomers. But the University of Pittsburgh project has a good chance of discovering several sets of planets, thus providing not only data for testing scientific theories but also a great boost to public enthusiasm for deep-space research. Because the cost involved is only about $200,000, private funding is quite feasible. Important contributions have come from the Extrasolar Planetary Foundation, established in 1980 especially to garner public support for such research.

Other projects to detect extrasolar planets are being carried forth by observatories and astronomers funded by more ordinary means. In August 1983, the Infrared Astronomical Satellite detected cool material in orbit around the star Vega, but the fact that Vega is a relatively young star means that the radiation may be from a large number of small objects, rather than from fully formed planets. This astronomical satellite, funded by the governments of Britain, the Netherlands, and the United States, is also capable of detecting previously undiscovered asteroids in our own solar system. Still, money collected by public subscription may be wasted if invested in private research projects that are not coordinated with research in the same fields conducted by NASA and the leading observatories. Indeed, public donations of only a few thousand dollars could provide the computer time and technicians' salaries required to complete analysis of vast quantities of untouched data from past space missions, going back to the Lunar Orbiter project of the 1960s. NASA prefers to spend its money on new missions, so private funds to complete analysis of already-collected data could help move other projects forward, as well as tidy up the loose ends of old research.

Public fund drives played a major role in building the great U.S. astronomical observatories of the nineteenth century. The 1856 report of the Harvard Observatory lists contributions from over a hundred of Boston's elite, sums ranging from $20 to $100,000. The Cincinnati Observatory was funded primarily by the middle class in a drive to which about 800 persons contributed. The August 1846 issue of *The Siderial Messenger,* a popular newsletter of astronomy, gave the occupations of 535 contributors. There were 45 in the legal professions, 29 in financial institutions, 30 teachers and clergymen, 41 physicians and druggists, 34 persons living from rents, 16 in publishing, and 117 miscellaneous merchants among those who donated cash; but there were also 43 in metalworking, 36 in woodworking, 9 masons and plasterers, 25 hardware and lumber merchants, and 44 miscellaneous makers and manufacturers, many of whom donated their labor and products rather than money. Thus, over a century ago, Americans in significant number were prepared to support astronomical projects through public subscription. What has been done before can be done again.

In recent years, government funding of modest projects to search the sky for radio signals from extraterrestrial civilizations has switched on and off unpredictably, but the public shows sustained interest.[11] Even mere detection of another civilization out across the stars, before any decipherment of the message, would shock a drowsy and unimaginative world into a wholly new perspective on the meaning of life and the proper future course of civilization. In 1983, with funding from the Planetary Society, physicist Paul Horowitz of Harvard embarked on a search using a 128,000-channel receiver he developed, that is attached to the university's 84-foot radio telescope. If another civilization is intentionally directing high-power radio signals at Earth, Horowitz has a chance of finding it. But his radio telescope is not sensitive enough to "eavesdrop" on the ordinary radio communications of even a nearby extraterrestrial society. Thus, his project may produce unjust discouragement if it fails to detect signals over the next few years; there is good reason to doubt the wisdom of insufficiently funded marginal searches which might ultimately do more to block interstellar contact than to achieve it.

Frank Drake of Cornell University has estimated that a seven-year program costing only $2 million a year stands a good chance of success using the most modern detection methods, but some of the projects that have been sketched would be far more expensive than this.[12] Through mass media programs like Carl Sagan's science television series *Cosmos,* and Steven Spielberg's films *Close Encounters* and *E.T.,* the public has been led to believe that extraterrestrial societies may not be difficult to find. But serious estimates of the number of habitable planets involve great uncertainties,[13] and the apparent absence of extraterrestrial visitors on Earth has led some scientists to conclude that humanity may be alone in the galaxy.[14] From the same facts, Schwartzman has derived the opinion that expensive attempts to communicate with extraterrestrials by radio are unnecessary because they are already watching us.[15] Furthermore, it is not clear that successful communication with an advanced civilization would be good for us. Space research might degenerate to a passive scholasticism, cataloging and interpreting the signals of extraterrestrials rather than also launching an aggressive program to explore and colonize the universe ourselves.[16]

Parallel Support from Science Fiction

Since before the beginning of the space flight movement, public enthusiasm has been encouraged by the parallel cultural movement, science fiction. Quantitative research I carried out on science fiction (SF) literature indicated that it really promulgates three competing ideologies about the future, only one of them enthusiastically supportive of the physical sciences and technology in general. But the field achieves a high level of consensus in favor of space travel. While the SF subculture is escapist, and does not directly launch effective campaigns for action, it instills favorable attitudes toward spaceflight and suggests that mankind can find among the stars a fulfillment of all the hopes and desires frustrated in the mundane world of ordinary society.

The youth counterculture that resisted militarism and promoted psychedelics in the 1960s has contributed a "hippie wing" to the movement. For a time, drug guru Timothy Leary expressed a poetic philosophy of outer space consciousness. A California organization that represents this uninhibited brand of activism is United For Our Expanded Space Program. Its initials of course spell out "UFO-ESP," and members describe themselves as "radical politically-active spacers." Among the other causes of this group is legalization of marijuana.

Far out from the military-industrial core of the original spaceflight movement is the fully cultic fringe, a large collection of flying saucer groups and extraterrestrial visitation cults which are an embarrassment to the rocket engineers and astronomers but which nonetheless may be of great social significance in the distant future. The most aggressive of these, Scientology, stands a good chance of becoming an influential, large denomination in the next century, following the trajectory of Mormonism, which was launched a hundred twenty years earlier.[17] At present, there are at least thirteen flying saucer religious cults in the United States,[18] but none seems destined for success. Ours is an age of religious innovation, however, and several utterly new religions of consequence will emerge from the tangle of little cults.[19] Perhaps Scientology will not be the only one of them to place conquest of the universe among its ultimate values.

The Military Provides a New Boost

Some of the most important fragments of the space flight movement operate unnamed and unseen inside the military-industrial complex. Among these is a strong "star wars" movement to develop high-power lasers and other beam weapons capable of knocking out reconnaissance satellites and perhaps even ICBMs. The magazine *Aviation Week* has continually urged development of orbiting battle stations since May 1977, when a long article warned that the Soviets' beam-weapon defenses against U.S. missiles would give them absolute freedom to undertake political expansion. A countermovement, seeking to halt development of beam weapons, finds its expression through the pages of *Science* and *Scientific American*. But without the right security clearance and connections, one can only guess what momentous decisions are being made behind closed doors.

Thus, the space flight movement may be on the verge of another successful *military detour*, as engineers interested in fulfilling their own career ambitions and personal dreams exploit the military's and political leaders' concerns about national security to the advantage of astronautics. The result may further world peace, this time, rather than war. Years ago, the Austrian space pioneer Eugen Saenger suggested that orbiting beam-weapon defenses might be the only means for ending the threat of nuclear destruction because they could stop attacking missiles or bombers without being capable of damaging civilian targets.[20] In March 1983, President Reagan announced his support of a similar idea, and in September an independent space group, High Frontier, established a political action committee to lobby for it.

One purpose of governments is to establish standardized means for settling disputes between powerful groups in society. But they are ill designed to deal with the great philosophical issues faced by the individual and by the human species, and cannot in the course of their ordinary operations plan and carry out cultural evolution to an entirely new level of existence. While radical social movements are extremely dangerous in this age of impending Armageddon, I think it is only they that can drive Earth to develop an interstellar civilization.

At present, the public space flight organizations are inoffensive and practically impotent. The profusion of activist groups does much to keep the universe in the public eye, and they may accomplish modest technical projects of some value to the future. Social conditions do not currently seem to exist which these groups could exploit to achieve further great leaps in space. Progressive militarization may produce the large orbital launch fleet required to support a system of beam-weapon battle stations. Colonization of the planets and exploration of the galaxy require the mobilization of extraordinary social forces. Barring some utterly unexpected technical breakthrough, conquest of the universe would require Earth to invest perhaps $100 trillion without a significant economic return. Only a mighty upheaval of the human spirit can accomplish this.

Toward an Interstellar Civilization

Interstellar civilization would represent a level of organization and culture immeasurably beyond anything yet achieved in our tiny planet. We may be incapable of imagining the stages of societal evolution that would bridge the immense gap from the present to that distant future. If interstellar civilization were easy, then the Earth would long since have been colonized from outside. But this does not appear to be the case. I suspect most intelligent species kill themselves or merely fade away, dying with a bang or a whimper. Maybe one in a thousand (or a million) bursts out of the confines of its solar system, driven by a radical religion, political movement, or internal conflict, to colonize a whole galaxy. Although the power to voyage across the stars implies the power of self-annihilation, a rare postindustrial society may somehow sail the tight course between quick immolation and slow decay.

We generally assume the current international balance of terror is a bad thing. The chance of sudden species death is all too real. But if we take seriously the conjecture that most technical civilizations that avoid violent suicide achieve the same thing more slowly through stasis, another perspective becomes plausible. Ours may be the best of all possible worlds, after all— or at least the best this side of Andromeda. Today's dire atomic threat may be historically necessary as a precondition for ultimate success. Perhaps the military phase of astronautics must proceed until large launch systems have been developed, and a transcendent social movement can take charge of colonization of the solar system, in turn providing the economic and technical base for interstellar travel. This must happen quickly, if the fall into stasis is to be avoided, so the course of history must run very close to Armageddon until the planets and their moons are won.

Ordinary bureaucratic policy will never take us to the stars. Perhaps a new religious denomination will appear, marching to the faith that the gods dwell somewhere across the universe waiting for us to visit them. Or perhaps the hope that will focus our energies skyward will be the belief that other civilizations have solved the problems which threaten to destroy us, and that they will give us guidance if only we can contact them.

The first phase of space progress was achieved by a social movement operating outside the ordinary institutions of society, but exploiting them whenever possible. Future revolutionary progress may follow the same course. In the end, the earthbound governments that currently set modest space policies may have to be transcended or abandoned. At present, the movement is biding its time, rallying public support and achieving small gains, waiting for those cataclysmic social conditions which might be exploited in a new rush forward.

Notes

1. Eric M. Jones, "Colonization of the Galaxy," *Icarus* 28 (1976): 421–22; Michael H. Hart, "An Explanation for the Absence of Extraterrestrials on Earth," *Quarterly Journal of the Royal Astronomical Society* 16 (1975): 128–35; J. N. Clarke, "Extraterrestrial Intelligence and Galactic Nuclear Activity," *Icarus* 46 (1981): 94–96.
2. Fred Hoyle, *Of Men and Galaxies* (Seattle: University of Washington Press, 1964).
3. Peter M. Molton, "On the Likelihood of a Human Interstellar Civilization," *Journal of the British Interplanetary Society* 31 (1978): 203–8; William Sims Bainbridge, "Computer Simulation of Cultural Drift: Limitations on Interstellar Colonization," *Journal of the British Interplanetary Society*, in press.
4. Sebastian von Hoerner, "Population Explosion and Interstellar Expansion," *Journal of the British Interplanetary Society* 28 (1975): 691–712.

5. Jose M. R. Delgado, *Physical Control of the Mind* (New York: Harper & Row, 1969).

6. Arthur C. Clarke, "The Best Is Yet to Come," *Time,* 16 July 1979, p. 27.

7. Vane R. Kane, "Interview with Bruce Murray," *Astronomy* 10 (September 1982): 24–28.

8. William Sims Bainbridge, "Collective Behavior and Social Movements," in *Sociology,* ed. Rodney Stark (Belmont, Calif.: Wadsworth, 1984).

9. Willy Ley, *Rockets, Missiles, and Men in Space* (New York: Signet, 1969).

10. Vernon Van Dyke, *Pride and Power: The Rationale of the Space Program* (Urbana: University of Illinois Press, 1964); Frank Gibney and George J. Feldman, *The Reluctant Space-Farers* (New York: New American Library, 1965); John M. Logsdon, *The Decision to Go to the Moon* (Cambridge: MIT Press, 1970); William Sims Bainbridge, "Public Support for the Space Program," *Astronautics and Aeronautics* 16 (June 1978): 60–61, 76.

11. William Sims Bainbridge, "Attitudes Towards Interstellar Communication: An Empirical Study," *Journal of the British Interplanetary Society* 36 (1983): 298–304.

12. Bernard M. Oliver et al., *Project Cyclops: A Design Study for Detecting Extraterrestrial Intelligent Life,* NASA CR-114445 (1972).

13. Alan Bond and Anthony R. Martin, "A Conservative Estimate of the Number of Habitable Planets in the Galaxy—Part 2," *Journal of the British Interplanetary Society* 33 (1980): 101–6; Anthony R. Martin, ed., *Project Daedalus: The Final Report of the British Interplanetary Society Starship Study.* Supplement to the *Journal of the British Interplanetary Society* (1978).

14. Frank J. Tipler, "Extraterrestrial Intelligent Beings Do Not Exist," *Quarterly Journal of the Royal Astronomical Society* 21 (1980): 267–81; J. N. Clarke, "Extraterrestrial Intelligence and Galactic Nuclear Activity," *Icarus* 46 (1981): 94–96.

15. David W. Schwartzman, "The Absence of Extraterrestrials on Earth and the Prospects for CETI," *Icarus* 32 (1977): 473–75.

16. Tong B. Tang, "Fermi Paradox and C.E.T.I.," *Journal of the British Interplanetary Society* 35 (1982): 236–40.

17. William Sims Bainbridge, "Religions for a Galactic Civilization," in *Science Fiction and Space Futures,* ed. Eugene M. Emme, pp. 187–201 (San Diego: American Astronautical Society, 1982).

18. J. Gordon Melton, *Encyclopedia of American Religions,* 2 vols. (Wilmington, N.C.: McGrath [A Consortium Book], 1978).

19. Rodney Stark and William Sims Bainbridge, *The Future of Religion* (Berkeley: University of California Press, 1984).

20. Eugen Saenger, *Raumfahrt—Technische Ueberwindung des Krieges* (Hamburg: Rowohlt, 1958).

13

The Social Forces behind Technological Change and Space Policymaking

James Everett Katz

How does technological change take place in societies, and what kind of challenge causes nations to decide to invest billions of dollars with little hope of receiving a penny in return? These two questions, which have received attention from sociologists, serve as the focus of the book *Space Flight Revolution* by William S. Bainbridge.[1] This chapter critiques Bainbridge's answer, and proposes an alternative to his view of the process of national space policymaking. While my comments are directed toward his complete book, a portion of his theory can be found in the preceding chapter.

The traditional explanation of technological innovation in modern societies is found in the work of W. F. Ogburn and S. C. Gilfillan, who located the driving force of technological change in larger social forces, especially economic forces, which manifest themselves through gradual, incremental processes of innovation and diffusion.[2] As J. M. Utterback has discovered, 60 to 80 percent of important innovations occur in response to market demands, that is, the "pull" of market forces, while the rest result from the "push" of scientific and technological advances.[3]

Bainbridge dispenses with the overwhelming bulk of innovation as "normal" technological change," basing his view of "normal" and "revolutionary" technological change on Thomas Kuhn's widely discussed book, *The Structure of Scientific Revolutions*.[4] While most technological change is "normal," Bainbridge concentrates on "revolutionary" technological change, or change that occurs outside conventional market mechanisms. The pursuit and attainment of manned space flight, he argues, constitutes revolutionary technological change in that it emanated from social movements and the visions of far-sighted individuals. Bainbridge holds that in order for revolutionary change to take place, a radical shift must occur in people's conception of what can be achieved and/or what it is desirable to achieve. This view of technological change (as opposed to Kuhn's tightly argued theories of scientific change) draws upon presociological "great-man" theories of history, which attribute change to individual personalities and deny the importance of links between established social roles and impersonal historical forces. Bainbridge examines revolutionary technological change from this viewpoint as it occurs through several generations.[5] The first, Generation 0, contains the intellectual precursors of change but does not generate new scientific insight. In Generation 1 are the intellectuals who lay the theoretical groundwork for later technological change through their correct conclusions about what can be done and how. Generation 2 is the key generation, containing the charismatic leaders who gain influence over enough resources and people to produce actual technological change or, in this case, space flight. Following Generation 2 is Generation N, the technicians who perform useful work within an already institutionalized (space flight) establishment, or the "normal" technological change generation, responding to market forces by refining an already established new technology.

Crucial to Generation 2 are historical circumstances that permit the exploitation of social conflicts so as to assemble the large programs and budgets necessary to foster a dramatic technological revolution. The wild ideas of Generation 0 have no tangible results; the abstract calculations of Generation 1 remain sterile treatises. It is only with the arrival of the exploitative and manipulative Generation 2 that the conventional institutions of society and the strains within power structures can be used to achieve massive change such as space flight. How does Generation 2 manage to push its programs through the crowded agenda of national decisions? According to Bainbridge, they have a little charisma, a lot of chutzpa, and, most of all, a rapport with the powerful and an ability to impress them and gain their support.

The revolutionary vigor of technological change is sustained, according to Bainbridge, by a social movement. He says that the space flight movement, for example, is supported by science fiction cults. This approach, however, ignores the vast amount of research on organizational processes and bureaucratic politics—where the key to setting and carrying out national agendas is invariably found.

Anatomy of a Revolution

It must be understood that the goal of the space flight revolution has never been simply to launch rockets into the emptiness around Earth. Rather, even from its inception, the space flight revolution has seen rocketry and missile development as an instrument to eventually put humankind into outer space.

Its grand design was to colonize space and the planets throughout the solar system and eventually the galaxy. Of course to prosaic people here on Earth such plans are simply the cloth of dreams. But those who believed in exploring the universe and sought ways of achieving it realized that such exploration must proceed in a series of increments.

Thus, to appreciate humankind's movement into space, it would be useful to examine the evolution of space colonization, for colonization has always been the ultimate goal of those space enthusiasts who have argued for rocketry and missile programs. By understanding the intellectual underpinnings of this movement, we can understand its several goals and the way it has attempted to shape national space policy.

Robert Salkeld has been most thorough in tracing the evolution of men living in free space.[6] He points out that, although visions of space travel are thousands of years old, the idea of space colonization is relatively recent. Newton mentioned the possibility in 1687, and several science fiction writers, including Hale, Verne, and Lasswitz, expanded it during the nineteenth century. Beginning in the early twentieth century, science fiction accounts of space colonization became more technically elaborate, such as the work of Konstantin Tsiolkovsky, whose description of a manned space station included rotation for artificial gravity, use of solar energy, and even a space "greenhouse" with a closed ecological system. Hermann Oberth elaborated on potential uses of space stations, noting that they could serve as platforms for scientific research, into the Earth and the cosmos. The concept of a rotating, wheel-shaped station was introduced in 1929 by Hermann Noordung, who suggested that it be placed in geosynchronous orbit.

By the time the Germans began military study of space stations during World War II, many of the basic ideas had already been proposed and elaborated. The main force in the German effort was Wernher von Braun, who after the war, popularized the idea of space colonization in the United States, where it was picked up by science fiction writers such as Arthur Clarke and Robert Heinlein and where it caught the imagination of scientists like Freeman Dyson and Fritz Zwicky.

Finally, in 1969 Gerard O'Neill led a group at Princeton into a technical investigation of the space colonization ideas that were proliferating everywhere, and they added a few more of their own which aimed at proving the possibility of a station rather than imagining an optimum one. O'Neill selected as a test case a rotating satellite using solar energy to sustain a closed ecological system. It was constructed of processed lunar ore delivered by electromagnetic accelerator, was located at Lagrangian point L-4 or L-5 to make delivery of lunar ore as simple as possible, and was configured as a 1-km-long cylinder with hemispherical end-caps and as having an internal Earth-like environment on the inner surface fed with sunlight reflected in by mirrors. To these elements, O'Neill added the following ideas:

- The habitat was configured as twin cylinders mechanically connected and contra-rotating to facilitate attitude control for solar orientation.
- The electromagnetic accelerator was adapted as a space propulsion device for transport between Earth-orbit and L-5, using lunar slag as reaction mass (mass driver).[7]

In a further step, the Princeton group under O'Neill recently suggested that the L-5 colony could construct satellite solar power stations (SSPS) from lunar material, and concluded that this would improve the economics of both the SSPS and the colony itself. The SSPS concept has received increasing attention since its introduction (and patenting) by Peter Glaser in 1968,[8] culminating in a 1974 conference, Space Manufacturing Facilities, sponsored jointly by Princeton, NASA, the NSF, and AIAA.

The conference marked the change from fictional visions to actual investigation of the technical feasibility of space colonies. About 130 registered participants from Canada, England, and the United States listened to 30 speakers discuss the technical, economic, organizational, and cultural challenges of establishing space colonies. Some participants were specialists from fields rarely represented in space conferences. The tone was serious, the technical content generally good, and the discussions lively. No problem emerged that would invalidate either the colony or the SSPS concept, though key technical challenges were identified. Perceptions naturally varied with regard to schedules and costs.

Since that first conference, articles have appeared about space colonization in such national magazines as *Newsweek*, *New York Times Magazine*, *Science*, *Physics Today*, and *Technology Review*, and O'Neill and others have held more conferences, some covered by PBS and BBC. The space colonization idea received considerable play in California's Space Appreciation Day celebration, and a large meeting to discuss the topic has been sponsored by numerous professional groups. In short, the permanent inhabitation of space has gained widespread acceptance as an idea and has been chosen by the U.S. government as the next major step after the Space Shuttle, although the time line is still indeterminate.

In its broad outlines the history of the space colonization movement follows Bainbridge's model, complete with dynamic leaders. However, below the level of public notice lies a complex network of organizational and institutional forces that determine the infrastructure of the space colonization movement and undergird U.S. national space policy. Not surprisingly, NASA is at the forefront of these forces, although various other organizations are also influential.

It is important that we distinguish here between superficial appearances of social change and the actual processes that affect social change. Bainbridge argues that a few individuals—leaders—are able to exercise control beyond their special area of concern and influence political actors in other arenas. He also sees the replacement of the "heroic" individual by faceless bureaucratic operatives once the revolutionary stage is completed. I have little quarrel with the idea that when a new area is exploited, entrepreneurial personalities are attracted to it. As organizational processes and technologies become well established and understood, the flexible, dynamic characteristics of the particular area are replaced by those of stability. Change becomes incremental rather than dramatic. There is certainly no question that personality interacts with social structure.

Where Bainbridge and I part is on the question of what drives the organization and the technological change. Bainbridge argues that the space man is able to manipulate the political man or the industrial man. My analysis suggests that the situation is actually the reverse. The main driving force is not the "great man" or the "genius" but rather how the social structure incorporates technological change. Technological innovations present opportunities that can be exploited by political, economic, and especially military systems. It is not very strange that a few individuals will be the innovators, the highly effective program managers, or even the slightly charismatic public figures. But these individuals can only modify technical programs or influence somewhat their perceived desirability. They cannot singlehandedly create a revolution to have the technology adopted. *This must come from an external desire to have what the technology can provide.* To illustrate my point, we can look at the National Aeronautics and Space Administration, NASA.

NASA as a Bureaucratic Entrepreneur

NASA was established in 1958 in the wake of the Soviets' stunning launch of Sputnik. It evolved out of the old National Advisory Committee for Aeronautics (NACA), which was established around the turn of the century to exploit the new technology of powered flight. NACA was not established because of visionary geniuses but, rather, because it made good organizational sense for the government to exploit technology that had important ramifications for a very large, thinly populated, transcontinental nation, as the United States was at that time. So too, it made perfect sense for NASA to be established in 1958 to exploit a new technology—artificial satellites and ballistic missiles—which had grave ramifications for a country that saw itself as the preeminent military, economic, and political power on Earth.[9]

NASA was not instigated by a few space men tricking those in power. It was formed by those in political power who wanted to exploit what the space men had to offer. Actually, they wanted a technology far beyond what the scientists and engineers were ready to provide in order to accomplish political and military objectives.

Today NASA's job is to run the nation's civilian space program. The military space program is controlled by the Defense Department. NASA has developed the Space Shuttle, which serves a joint civilian–military role. It has also developed weather and other remote sensing satellites and conducts astronomical and atmospheric research, to name a few of its many activities. On the one hand, these are relatively bureaucratic operations, that is, "ordinary," in Kuhnian terms. But now NASA, with the White House's blessing, is moving ahead with the next step on the "revolutionary" road to space flight: a permanent manned space station, the first step to a space colony. This is indeed a revolutionary activity, but it is a revolution without a space man duping a power man for his own purposes. Rather, it is technology being harnessed by an organization in the service of politically and militarily defined needs. Thus the preeminent leadership on space development revolves around not a few individuals but a formal institution. NASA, its industrial contractors, its supporters in the executive branch (including the White House, and Defense Department), and Congress are successfully pressing for a vigorous space policy—one that will include a space station.

In one sense, the ambitions of space advocates simply coincided with the political exigencies of the day. The Red Menace fear practically drove the United States to adopt a vigorous space policy. Had the Soviets initiated a deep-sea drilling project—in an attempt to get to the center of the Earth or drill deeper than any other country had done before—it is very likely that the United States would have followed suit. By the same token, if the Soviet Union begins landing people on the moon and developing colonies there, the U.S. space program would quickly accelerate. It is the Soviet competition that has sent the United States out into space, not the Earthbound space man. Likewise, it is only Soviet competition which is bringing life back to the U.S. space program today under the Reagan administration, and which will put a space station in orbit.

While it would be possible to argue that the visionary space men are the ones who patiently prepared the way for the space man-leader to exploit the politicians' interest in space policy, one would have to make the same argument for submarine enthusiasts, dirigible enthusiasts, deep mining enthusiasts, UFO enthusiasts, and so on. It has been the military applications only that have brought space exploration to the top of the national agenda.

Leadership is important, but it can take effect only through an organization and within the context of decisions to allocate resources. Social processes and roles become pivotal in creating and implementing policy. By examining NASA's influence over the formulation of space-settlement policy, we can see these processes operating in detail.

NASA's Role in Space Settlement

The potential direction of the space program is designed in NASA's long-range planning office, which is responsible for detailed five-year plans for the program's operations. Ten- and twenty-year plans are also drawn up and meshed with the five-year plans to provide policy planners with an outline indicating the ultimate direction of the national space program. Although some consideration is also given to a period farther into the future, the view beyond ten years is uncertain. The long-range planning office establishes the space program's goals and determines the resources available for them by estimating the total NASA budget in ten years, and subtracting the resources that the current programs will require. The amount left over is called the "planning wedge." In choosing the goals with which to orient the planning wedge, there is a continual interplay between what can be done, what should be done, and, therefore, what is done.

Understandably, and probably justifiably, NASA hopes that interest in space and space budgets will increase in the years to come. The 1960s are fondly remembered for their excitement and big budgets, but NASA found itself seriously constricted after Apollo. Public interest wavered and there was neither competition from the Soviets nor a dramatic space spectacular available to bolster the agency's sagging fortunes. The future constituted a sharp challenge for NASA leaders. Because energy was a hot topic even before 1973, NASA considered adding "E" for energy to its title and becoming the national energy and space administration. This did not become a real interest, however, because NASA officials believe both the agency's and humankind's destiny lies in outer space. They assume, a priori, that space exploration benefits contemporary society, that humans must conquer space and extend their dominion into the solar system and beyond. This assumption results from what has been called the extraterrestrial imperative. It will remain the basis for NASA's objectives, despite NASA's dabblings in energy, urban technology, and other side roads.

Thus it was that space colonization emerged as an exciting new possiblity to engage NASA and stimulate public support. Unlike lunar or Martian colonies, settlements in space are feasible enough to be realized quickly and are potentially of immense benefit to Earth. NASA would be serving itself while serving humanity by spearheading the space settlement movement. Also, colonization is the next logical step after the Space Shuttle begins regular service and large space facilities such as telescopes, power stations, laboratories, and manufacturing installations are developed. In this way NASA will continue to serve an important function and may thereby insure its survival.

Because of a mutuality of interests, NASA has supported the space settlement movement. Since 1975, annual summer study groups, consisting of noted or potentially important academic figures, have been convened to investigate various aspects of space settlement. High-level specialists within NASA and the aerospace industry have served as their consultants. Through this effort, much of the technical groundwork can be laid and major problems addressed. NASA has also used its in–house expertise to study space settlement problems. In fact, even the name "space settlement" was chosen by NASA, when the original name, "space colony," was objected to by less developed countries as being reminiscent of an imperialistic era; they feared a movement by the advanced nations to re-establish colonial relationships. The term "space habitat" was also considered but rejected as too sterile and antihumanistic. "Space settlement" was selected because it evoked the pioneering spirit at the same time that it suggested "hearth and home." Thus, even the title of the movement was chosen only after thorough assessment by NASA, a triumph of manipulated image over spontaneous imagination.

NASA's advance planning office has been seeking funding for about a hundred dissertations on space settlements over the next several years, allotting $5,000 to $10,000 to each. These grants will support research on problems of space colonization, and build up throughout the nation an academic cadre of scientists and engineers with an interest in deep space who will form a research base dedicated to developing space settlements. Each professor should in turn produce new students who are committed to space settlements research.

The long-range planning office has been anxious to aid the large number of supporters of space settlement who now constitute a social movement. The most technically and programmatically useful work NASA can perform is to continue its present function: operating and expanding activity in space. By extrapolating Space Shuttle work and by building orbiting space stations, NASA will be enhancing the nation's space capability and at the same time increasing the "sunk costs" and elevating the potential efficiency of large, permanent space facilities. Such activities add to the possibility of the space settlements' being placed on the national agenda. The heroic figures have disappeared and been replaced by an organization like any other in its focus to perpetuate itself. Its leading personalities are bureaucratic, not individualistic.

NASA is not alone in supporting activities related to space settlements. Professional groups such as IEEE and AIAA have underwritten conferences on the subject. SRI, a nonprofit think tank, and the British Interplanetary Society, among others have contributed to advancing the cause, as have foundations such as the Marlar and Lindbergh. MIT, Princeton, and similar universities allow their facilities to be used to advance space settlement planning, not through any conscious decision by their boards or directors but as a result of decisions of students and faculty.

These organizations and NASA prefer to operate off center stage as supporters of, rather than as visible evangelists for, space settlements. The evangelical role is played by myriad individuals from academe, private industry, and the interested public. Because the evangelism is amazingly widespread and draws upon both popularizers and researchers, space settlement has become an ideology to believe in as well as a plan to act on.

Space Settlement as a Social Movement

The space settlement idea has assumed the proportions of a social movement and fits within such definition. It features collective behavior and action to promote change, and exhibits the characteristics of shared values, a goal sustained by an ideology, a sense of membership, norms, and a structure facilitating divison of labor. It has grown from cumulative cultural change and a general increase in space exploration and technical capability. Its advocates perceive themselves as constituting the wave of the future, and are developing a program to hasten and direct changes and to rally those of a similar persuasion to promote change that will affect human destiny. They see the alternative as acquiescence to becoming passive subjects of cultural drift. The space settlement movement combines the means of personal participation, personal transformation, and societal manipulation.

The movement has split into groups that might be designated the "hard-headed realists" and the "fuzzy dreamers," although the distinction is more one of style than of content. The hard-headed group, led by Gerard O'Neill, a Princeton physicist, is composed of scientists and engineers who devote their time to developing the technical designs to implement the space settlement ideas. The dreamers operate largely through the L-5 Society, and are concerned with creating and maintaining public attention and enthusiasm for usefully deploying humankind in space. Carolyn Henson, owner of an electronics firm in Tuscon, coordinates this wing of the space settlement movement. A variety of other groups, each with its own newsletter and journal, have sprung up to link, sustain, and reinforce their members.

While the scientific group was once identified with wild-eyed radicalism, it was quick to assume a posture of "technocratic realism." O'Neill sees 1975 as the turning point: "A year earlier, we had been a small, happy band of revolutionaries; now with increasing recognition by professional bodies, it was desirable and necessary to adopt a conservative and pragmatic approach."[10]

The popularly oriented L-5 Society has concentrated on seeking legitimacy through symbol manipulation. Members have appeared on radio shows, published magazines and booklets, and held meetings. The society has fifteen chapters nationwide and about eight thousand members. Membership fees, gifts, and contributions support several full-time staff members, a large number of volunteer workers, and a slick monthly magazine, the *L-5 News*.

Both these groups have well-established communication networks and they attempt to lobby Congress. There is some tension between the technocratic and popular wings of the space settlement movement, but they do cooperate. If the technocratic wing separates too far from the popular wing, it risks losing its support, while the popular movement is in danger of becoming so extreme or dogmatic that it will arrive at the outer fringes of legitimacy. Each needs the other to sustain influence and viability.

It is easy to imagine likely scenarios for the space settlement movement, but it is difficult to predict which of them will materialize. Each vision of the future holds a different prospect for the structural position of the space settlement movement. Which one will be realized depends on the internal or external societal pressures that will enlist public and governmental support for space settlement. The competition from the Soviets, the need for more resources, especially energy, or the movement itself may be enough to establish space settlements. It is possible that no extraterrestrial imperative exists and, if so, the space settlement movement could be shunted aside as the geocentrists or Lamarckists once were, and humankind would remain Earthbound. It is also possible that humans will expand into space through a format other than space settlement, causing the space settlement movement to be bypassed, much as balloonists were eclipsed by aviators. There are still dirigible advocates and groups but they find little support. It is also possible that the space settlement movement will be wildly successful, surpassing all expectations. In this case, its leaders would stand at the king's elbow, so to speak, whispering directives about establishing colonies. Alternatively under this scenario, the movement would be co-opted and absorbed into the mainstream culture, replaced by bureaucratic experts. The most probable sequence is limited and gradual success. Enthusiasts would then assume the public side of a political lobby, similar to the environmental or national security interest groups. The eventual success of the space settlement movement could dramatically change our lives, but the possibility also exists that space settlement would permit us to remain the same—to consume plenty of energy, to use lots of natural resources, and to continue to multiply our numbers. (Most scenarios project today's values into the future, anticipating a veritable "Disneyland utopia.")

Organizing and Managing Technological Change

It is now appropriate to return to the issue raised at the beginning of this paper. I contend that although technological change is set in motion by individuals, the proper social and institutional, not to mention technical, environment is also necessary for its initiation. In contrast to Bainbridge's theory of revolutionary change initiated by individuals, studies of government decision–making reveal bureaucratic politics and institutional imperatives as the engines of large-scale technological change. A case in point: space settlement has everything in place, including technological capability, to succeed, yet it lacks the broad-based cultural and political support needed for success, not because of a lack of leaders but because the potential benefits are too uncertain in view of the obvious costs.

Lacking any external pressures such as Soviet competition or a perceived exhaustion of energy resources, bureaucratic strategies are necessary to protect a program once it has been born. Proponents of any program, including space settlement, need skill in bureaucratic politics and an understanding of the political nature of technological change if they are to succeed.

Research on technology programs identifies three major survival strategies: differentiation, co-optation, and moderation.[11] It is through differentiation that organizations establish unchallengeable claims on resources by distinguishing their product from those of competitors. Co-optation involves absorbing new elements into the leadership or new ideas into the program to defuse threats to a program's stability. Moderation is the ability to forgo short-term gain to build long-term support.

Within programs, movements, or bureaucracies there are certain key roles that help bring a technological project to fruition. These roles or "critical functions" also are necessary to shepherd a program toward fulfillment and to overcome threats to its existence.[12] The first of these functions to occur and the most obvious is idea generation,[13] the act of invention. It is performed by people who can sufficiently remove themselves from the routine and the obvious to see the less obvious, and who have the perseverance it takes to establish technical feasibility. They are the creative scientists or engineers who excel at conceptual thinking and analyzing situations. The second critical function is that of the entrepreneur,[14] who pushes the new idea forward from technical feasibility to a potential practical application. Entrepreneurs attract the attention of the relevant public by advocating the new technology as a bigger, better, cheaper, cleaner, or whatever way of doing something. They argue for change, and their talent lies in transforming the technical breakthrough of the creative scientist or engineer into a marketable product or process.

The project manager[15] usually takes over after the project has been established as a legitimate challenge to the present way of doing things and has some money behind it. The task is to organize the diverse sets of activities (technical, market, financial, etc.) that must take place to move the product toward the market or receiving unit. The project manager's skills are in planning, organizing, and administering.

Three further roles have been found critical to support the creative scientist or engineer, entrepreneur, and project manager: the technical information gatekeeper, the market information gatekeeper, and the sponsor. The first provides the expert technical information about what is occurring in a specific technical field by reading journals, attending conferences, and staying in touch with colleagues doing related work.[16] Equally important, technical information gatekeepers channel this information beyond their own needs to a network of associates for whom they are important contacts with the "outside world"; they become invaluable to technical problem solving. Their skills are their ability to communicate.

Performing a parallel function is the market information gatekeeper, who serves a crucial role in defining problems and useful solutions. Much as the technical information gatekeepers stay abreast of research and scientific publications, market information gatekeepers remain aware of market trends, customer reactions, news from suppliers, and changes in the social or political environment. They channel their data to colleagues to help with planning and problem solving, and are thus also talented communicators.

The sixth critical function is that of the sponsor,[17] one who assembles the resources for the project, makes connections through his or her network of contacts to keep the project moving ahead, and provides valuable advice (technical and administrative) to the people working on the project. Usually they are senior persons who may also help the project people set goals and assess progress.

All six of these functions are critical and thus require that a range of people with very different types of skills and sensitivities be associated with a project. Teamwork throughout the course of a project is necessary, with the centrality of roles shifting over time from inception to application. There is also a reward system that recognizes very different types of contributions as essential. Most important, the critical functions concept underlines the point that neither market nor technical information is sufficient, though both are necessary, for utilization and eventual success. An admixture of key people who are able to make appropriate use of information must be associated with the work to insure its success.[18]

These six functions and the three strategies discussed earlier negate the idea of revolutionary technological change. Ideas and programs are realized through set principles, which are sufficient to explain all technological change and programmatic success without the addition of an elegant but inappropriate theory like Bainbridge's. Technological innovation can be better understood as a result of interacting roles and organizational imperatives than as the product of intergenerational change.

It can thus be argued that national space policy is set by the same pressures that determine national security, foreign, and farm policy. Even space colonization does not represent revolutionary technological change. A product must be sold in the marketplace—public, governmental, or private—whether the product is an idea, a guided missile, a hair brush, or a colony in space, using the same strategies and under the same limitations, to wit: Every technology is preceded by feckless dreamers and visionaries. Every idea is "sold" to someone else—either before its design, when research or production support is needed, or after, when markets are needed.

Just because one perceives a series of "generations" in the development of a technology does not mean that one can use those generations to adequately describe and understand the causes of dramatic technological change. That is somewhat like saying, "I see that it gets dark before a rainstorm; darkness must be the cause of rain." Technological innovation and change are much more complex phenomena than can be described by the thoughts and actions of a handful of people. The ideas of paradigms and intellectual revolutions that Bainbridge relies upon remain more properly in the field of the philosophy and sociology of science. Techniques, tasks, and technology are governed by other factors.

Notes

1. William S. Bainbridge, *The Space-Flight Revolution: A Sociological Study* (New York: Wiley, 1976).
2. W. F. Ogburn, *Social Change* (Magnolia, Mass.: Peter Smith, 1964; reprint of 1950 volume); O. D. Duncan, ed., *William F. Ogburn on Culture and Social Change* (Chicago: University of Chicago Press, 1964); S. C. Gilfillan, *Sociology of Invention* (Cambridge: MIT Press, 1970).
3. See James M. Utterback and Chris Hill, *Technological Innovation for a Dynamic Economy* (Elmsford, N.Y.: Pergamon, 1979); James M. Utterback and William J. Abernathy, "Dynamic Model of Process and Product Innovation," *Omega: International Journal of Management Science* 3, no. 6: 639–56; James M. Utterback, "Innovation in Industry and Diffusion of Technology," *Science* (15 February 1974): 620–26.
4. Thomas Kuhn, *The Structure of Scientific Revolutions* (Chicago: University of Chicago Press, 1970).
5. Bainbridge, *Space Flight Revolution*, Ch. 2.
6. Robert Salkeld, "A History of Space Colonies," *Aeronautics and Astronautics* 14 (Spring 1976): 81–93.

7. Ibid.; Gerard K. O'Neill, *The High Frontier: Human Colonies in Space* (New York: Morrow, 1976); "Low Road to Space Manufacturing," *Astronautics and Aeronautics,* 16 (March 1978): 23–32; Study Director, *Space Resources and Space Settlements,* NASA Technical Paper SP-428 (Washington, D.C.: NASA, 1979). There have been numerous studies of space colonization; see chapter 1 in this volume.

8. Peter Glaser, "First Steps to the Solar Power Satellite," *IEEE Spectrum* 16 (May 1979): 52–58.

9. James Everett Katz, *Presidential Politics and Science Policy* (New York: Praeger, 1978).

10. O'Neill, *High Frontier*, p. 256.

11. Harvey Sapolsky, *The Polaris System Development: Bureaucratic and Programmatic Success in Government* (Cambridge: Harvard University Press, 1972).

12. T. J. Allen, "Critical Functions in Innovation," unpublished paper, MIT School of Management, Cambridge; M. Jelinek, *Institutionalizing Innovation* (New York: Praeger, 1979); Thomas J. Allen, *Managing the Flow of Technology* (Cambridge: MIT Press, 1977).

13. D. Pelz and F. Andrews, *Scientists in Organizations* (New York: Wiley, 1965).

14. E. Roberts and A. Frohman, "Internal Entrepreneurship: Strategy for Growth," *Business Quarterly* (Spring 1972): 71–78.

15. I. M. Rubin and W. Seelig, "Experience as a Factor in the Selection and Performance of a Project Manager," *IEEE Transactions on Engineering Management* EM-14 (September 1967): 133–35.

16. T. J. Allen, "Communications in the Research and Development Laboratory," *Technology Review* 70 (October-November 1967): 21–35.

17. E. G. Roberts, "A Study of Innovators: How to Keep and Capitalize on Their Talents," *Research Management* 11, no. 4 (1968): 249–66.

18. A. Frohman, "Critical Functions for an Innovative R & D Organization," *Business Quarterly* (Winter 1974): 72–81.

14

Extraterrestrial Intelligence: The Social Impact of an Idea

Ron Westrum, David Swift, and *David Stupple*

With infinite matter available, infinite space,
And infinite lack of any interference,
Things certainly ought to happen. If we have
More seeds, right now, than any man can count,
More than all men of all times past could reckon,
And if we have, in infinite nature, the same power
To cast them anywhere at all, as once
They were cast here together, let's admit—
We really have to—there are other worlds,
More than one race of men, and many kinds
Of animal generations.

—*Lucretius,* The Way Things Are[1]

The idea of a plurality of inhabited worlds is indeed an old one, having inspired and troubled humans since ancient times. Whether humankind is alone in the universe, or whether there are other worlds with intelligent beings on them, possibly superior to us in knowledge, are questions that have long exercised our reason and our imagination. With the development of modern space technology and radio astronomy, we have taken the first steps to explore nearby worlds directly and to scan far-off ones for signals from intelligent life. Our civilization is now at the stage where we have not only evolved theories about extraterrestrial intelligence (ETI) but also developed means for detecting and communicating with such intelligence. Yet the technical feasibility of detection and contact must confront broader questions of social policy. How much money is to be spent on such searches? Who will be responsible for carrying them out? What kind of searches should be carried out? To date, no searches for electromagnetic signals have been positive. Are further searches worthwhile? What kind of attention, if any, should be given to investigating UFO sightings, whose reported strangeness seems to suggest extraterrestrial origins?

Whatever decisions we as a society make on these issues in the long run, the concept of ETI is affecting our society in many ways—through science fiction books, films, and television programs; with respect to our religious beliefs; through public interest in the space program; and in regard to our self-image as unique beings in the universe. Our children play with toy rocket ships and models of "E.T." This response to the idea of ETI strongly reflects the hopes and fears of our society about its own future welfare.

Our purpose here is to examine American society's response to the ETI concept. To do so we have chosen three subcultures in American society whose existence is based on the idea of extraterrestrial contact. The first subculture consists of *religious groups* who believe that contact is now taking place and that messages from "space brothers" have already been given to human beings. The second is made up of groups and individuals who investigate UFO sightings, the *"UFOlogists."* The third subculture consists of *scientists* who are actively involved in research and thinking about intercepting signals from extraterrestrial intelligent species. Each of these subcultures believes that ETI exists and that it is making active efforts to contact us. Apart from this assumption, though, the membership, social structure, norms, and intellectual underpinnings of these groups are quite different. Each tends to regard the others with indifference or even scorn, in part because of differences in style, methods, and membership, but also because they represent three competing responses to the same basic idea.

Space Brothers Speak

If there are people on other planets, then why can't they come here? And if they do come here, then why don't they land on the White House lawn? Might they pick a humble person to communicate with? After all didn't God pick Jonah to deliver one of his messages?

This basic premise, that extraterrestrials could arrive on earth and communicate with ordinary people, underlies the entire "contactee" subculture we are about to examine. It provides a rationale for individuals who have had "contacts" to speak publicly about their experiences, for others to elaborate doctrines based on such experiences, and for the formation of organizations to collect and disseminate such testimonies and doctrines. The inherent interest in such testimonies (if true) and the vividness with which they are related also allow for their exploitation by the mass media—particularly by radio and television talk shows. No matter how implausible the basic premise might seem to others, there is no question that for millions of Americans this premise not only makes sense but seems to have been proved correct by the testimonies contactees gave about their meetings with the "Martians," "Venusians," or whatever. Today, there are hundreds of small religious groups in the United States who believe that extraterrestrials are in contact with Earthlings and that they wish to help us.

George Adamski was the first of the contactees. In 1952, at age 61, Adamski claimed that he had a meeting with Orthon, a man from Venus, near Desert Center, California. Before his experience, Adamski had been a handyman and an undistinguished teacher of "metaphysics." He was also interested in photography and had previously produced pictures of UFOs, later labeled fraudulent by the Air Force. In 1953, in a book published with the British writer Desmond Leslie, *Flying Saucers Have Landed,* Adamski gave the first of many accounts that he and others were to offer of meetings with extraterrestrials. Other contactees, among them Truman Bethurum, Orfeo Angelucci, Howard Menger, and Daniel Fry, followed quickly, giving their own accounts of such meetings and of the behavior and message of the "space brothers." How sincere they were is open to question. Adamski, Menger, and Fry were caught in falsehoods, and others were very suspect, but their "messages" had effect.

The contactees were a predictable attraction for radio and television talk shows; Long John Nebel, Betty White, and Steve Allen all featured them. The 1950s were a time when the threat of atomic war seemed very real. For many people, it must have been comforting to believe that benevolent intelligences from other planets were watching over them and that they were communicating messages of warning and hope to Earthlings who would then pass them on. Furthermore, such accounts were intrinsically exciting, if improbable. The media found in the contactees an easy way to exploit the high public interest in flying saucers. The contactees themselves basked in the attention, and proceeded to translate it into invitations to speak at "platform societies" and to found their own "flying saucer clubs."

Platform societies sponsor a succession of speakers on topics relating to psychic abilities, religion, spiritual healing, mysticism, and similar matters. Such groups sometimes sponsor psychic fairs, where local practitioners exhibit their skills in astrology, tarot reading, and various practical parapsychologies. They also may offer courses in the "mystic arts." The members of platform societies, contrary to an influential study published in 1966, are *not* society's marginals, the elderly, infirm, or unadjusted who "cannot carry on a conversation." Religious scholar J. Gordon Melton, familiar with many of these groups, suggests that their members are "ordinary people with some extraordinary beliefs.'' Although many of them are well educated, they tend to be uncritical. They accept each speaker as having, in their words, ''part of the truth.'' In Detroit, a common metaphor compares rival esoteric systems to looking at a house through different windows; each has its own view of the same room. The contactees, whose stories of exotic contacts with ''space brothers'' fit easily into the framework of these groups, became popular speakers for them. The attention they received, moreover, allowed the contactees to develop their own groups, and many of their disciples learned useful organizational techniques from participation in the platform societies.

From about 1954 to 1959, the flying saucer clubs became a major vehicle for the organization of people interested in the religious aspect of extraterrestrial contacts, and they were reinforced by a succession of newsletters, magazines, and articles devoted to flying saucers and their ostensible occupants. By the late fifties, the contactees' popularity was eclipsed by the growth of the publishing industry that had developed to promote them and the idea of flying saucers. *Fate,* a popular magazine devoted to the occult, found that publishing flying saucer stories helped its circulation immensely. One of its editors, Ray Palmer, left *Fate* and founded his own magazine, *Flying Saucers,* which was strongly dependent on the contactee and ''UFOlogy'' subcultures. Another publisher, Gary Barker, got his start in the business through a ''creature story'' published in *Fate,* and he went on to write *They Knew Too Much about Flying Saucers* (1956), the book that first described the ''three men in black'' who threatened witnesses. The men in black were to become a staple item of UFO lore. Barker eventually set up the Saucerian Press, which continues to publish books oriented to contactees and the more exotic aspects of UFO folklore. Barker's mailing list now includes 4,500 names, mostly people oriented to contactee themes rather than to occult or metaphysical ones, according to evidence from a reader poll. As the popularity of the better-known contactees waned, these publishers became celebrities themselves, often in demand on the lecture and convention circuit.

But the 1960s witnessed the decline of the flying saucer clubs and the fragmentation of the contactee subculture into a large number of only loosely related segments. The most easily identifiable groups are relatively large and organized charismatic congregations, such as George King's Aetherius Society, and are dependent on a single figure for leadership and prophetic messages. The Aetherius Society has perhaps 1,500 members in the United States and England. On a similar scale are psychic groups, such as Ruth Norman's Unarius Foundation in California, that often have substantial mailing lists and for which extraterrestrial contact is an important topic of interest. Nonetheless, there is a larger number of small groups whose members are also interested in hearing about messages from space. These include the platform societies as well as smaller groups organized for discussion, reading, and sometimes projects related to contact ideas. One such group was building a large "flying saucer" that, it was believed, would one day fly. Several highly skilled members of the group worked on making the vehicle structurally sound (the propulsion unit was to be supplied by the extraterrestrials). The group seemed to exhibit a working-class culture of recreational excitement, and much of the group's attention was focused on the physical task of getting the saucer built.

These diverse contactee groups also share several literatures, including books on contact experiences, astrology, psychic science, "metaphysics," nontraditional healing, and various forms of self-help. Much of the contactee literature has a "new age" emphasis, in many ways paralleling the Age of Aquarius theme in the youth movements of the 1960s. Much of it contains a millennial message suggesting that the space brothers are soon going to intervene and help mankind in various ways. Much of it, in other words, contains "good news for Spaceship Earth" from the extraterrestrials, who are usually described as wise and benevolent. The literature also contains a contactee folklore, mostly manufactured by the writers and publishers. The literature, its personalities, and its folklore provide the thread that ties together (very loosely) the contactee subculture.

In sum, the contactee subculture, born in the 1950s, is now maintained through an extremely diverse set of groups, some of which are organized as religious sects or as contactee cult groups. Others are set up as study groups or around a project, such as the saucer-building group mentioned above. These groups borrow skills and ideas from nontraditional religious systems (such as theosophy) as well as from psychic societies; they also draw upon UFOlogy, though to a lesser extent than one might expect. The members of these groups are largely working class or lower middle class and tend to have a lower educational level than UFOlogists, to whom we will next turn our attention. The religious needs to which these groups appeal and the opportunity for sociation in the doctrinally tolerant context they provide will probably make the contactee subculture viable for many years to come.

Collective Behavior or Protoscience?

UFO sightings, and those who investigate them, can be viewed in either of two very different ways. On the one hand, the phenomenon might be seen as an example of collective behavior and the contagion of perceptual experiences, and one might seek the supposed "structural strains" that led to the outbreak of such experiences in American society. Or it might be seen as the courageous struggle of inquisitive amateurs to bring to light events that the scientific community would just as soon ignore. Both views have some substance, and both can be supported by emphasis on one or another portion of the social behavior exhibited by sighters, investigators, the mass media, the military, and the scientific community (to name only a few of the actors involved). The treatment of UFO experiences by American society is complex, and here we can sketch only its outlines.

While experiences that might be labeled "UFO" go far back in history, the concept "flying saucer" was first applied to an actual experience in June 1947, after a spectacular sighting near Mount Rainier led to extensive newspaper coverage. Within weeks, hundreds of sightings had been reported in newspapers all over the United States. One sociologist, Herbert Hackett, was convinced that the flying saucer "craze" was "almost wholly a manufactured concept" and that it would quickly disappear. This expectation, however, was incorrect. The vividness and popularity of the "flying saucer" (a.k.a. "unidentified flying object") concept meant that it could and would be applied to a large number of aerial objects—some of them, to be sure, identifiable enough. In the 1960s, public opinion polls showed that about one in twenty adults believed that he/she had seen a UFO; by the 1970s, this figure had risen to one in ten. Not only did UFO sightings not go away, they continued in full force, although their apparent extent has varied with shifting media coverage.

Interestingly enough, UFO sighters are roughly a cross section of the U.S. population. They are not distinguished from nonsighters by income, education, race, sex, or religion. Although one study did suggest that some of them might be status-inconsistents, a follow-up study did not support this hypothesis. There is, however, one factor that is consistently related to UFO sightings, and that is youthfulness. The data show that the younger one is, the more likely one will report having sighted a UFO. While the meaning of this finding is not clear (it may have to do with differential willingness to label a perception a "UFO"), it suggests that there is some relationship between social structure and UFO observations. In any event, there is a strong predisposition for people to perceive UFOs, as indicated by the ability of UFOlogists to identify about 90 percent of reported sightings as known objects. UFOs, then, are part of our culture—and of the cultures of other nations, as well, since UFO sightings are a worldwide phenomenon.

The intense interest that UFO sightings evoke, of course, has to do with the possibility that some of the observed objects might be extraterrestrial vehicles. This possibility, and others, led to the early investigation of sightings by the U.S. armed services, notably the U.S. Air Force. In the twenty-year period that the Air Force was publicly involved in the investigation of UFO sightings, it sponsored two major studies—one by the Battelle Memorial Institute, the other by the University of Colorado. The overall conclusion of both studies was that UFOs did not provide evidence of extraterrestrial visitation, although the data were actually far from unequivocal. In both cases, there is some evidence of analysis being slanted to please the sponsors of the research. The Air Force's own UFO investigation unit, variously called Sign, Grudge, and Bluebook (reflecting its shifting orientation), was under even more pressure to come up with negative conclusions, and usually did.

Continuing publicity given to UFO sightings by the mass media, magazines, and books made some of the public impatient with Air Force investigations, which seemed slow, uninspired, and perhaps dishonest. The latter impression was reinforced by the books of a retired Marine Corps major, Donald Keyhoe, who used contacts inside the Air Force to find out ''what was really going on.'' The combination of curiosity and frustration led civilians to found their own organizations to investigate sightings. The Aerial Phenomena Research Organization (APRO) and Civilian Saucer Intelligence (CSI) were formed in 1952; and the organization destined to become the largest, the National Investigations Committee on Aerial Phenomenon (NICAP), was created by Keyhoe and others in 1956. At its peak, the now defunct NICAP had 10,000 members and published several large compilations of information on sightings. The largest and most active organization currently is the Mutual UFO Network (MUFON), which has about 1,500 members and an investigator network covering the United States. MUFON's annual conventions bring together many of the intellectual leaders of the UFO movement.

UFO investigation organizations display many of the usual problems of voluntary associations. Success is often dependent on leadership; finances are a recurrent problem; ensuring membership participation and conformity requires strenuous effort; and factions are common. The organizations would like to present their work as scientific, but scientists are a tiny minority among UFOlogists. Typical of the problems faced is the constant struggle to get members to submit investigation reports on standardized forms, or to submit any reports at all. In spite of these difficulties, a fairly large volume of investigation (of varying quality) does get carried out, and many of the more interesting reports are published in UFO periodicals, conference reports, and books.

What is the nature of the knowledge so developed, and what is its value? This is very difficult to assess. Investigation of UFO sightings is carried out ordinarily as a hobby in one's spare time, on weekends, evenings, or on time bootlegged from one's job. The cost of investigating an interesting case probably approximates that of a difficult homicide investigation; the time and money for a thorough job are seldom available. The number of interesting cases of close encounters, landings, physical-trace cases, sightings of ''humanoids,'' even ''abductions'' in the literature is quite large. It is as difficult for most insiders *not* to believe that something unusual is taking place as it is for outsiders to believe something is. Certainly the extent and strangeness of the reported events seem to call for continuing inquiry. Nor is it correct to believe that when scientific investigators look at the same cases, they find nothing. Both the Battelle Memorial Institute study and the University of Colorado study (popularly known as the Condon Report) had a large fraction of unexplained cases from "reliable witnesses."

UFOs have not been entirely ignored by other scientists. In 1953, Donald Menzel, a distinguished astrophysicist at Harvard, published *Flying Saucers*, a substantial book whose thesis was that sightings are largely attributable to aerial mirages. Although his specific optical explanations were later discounted, Menzel's authority was probably quite influential in persuading his colleagues not to take UFOs too seriously. (Menzel's coauthored second and third books appeared in 1963 and 1977.) In 1965, a different approach was taken by the astrophysicist and computer scientist Jacques Vallee, whose *Anatomy of a Phenomenon* turned Menzel on his head. Menzel had argued that people were good at seeing what was not there; Vallee argued that society was good at not seeing what was there, and that the UFO reports and science's response to them were well within the parameters for a real extraterrestrial visitation. Vallee's thesis received additional support in 1972 with the publication of J. Allen Hynek's *UFO Experience*.

Hynek's book was the first truly scientific examination of the UFO phenomenon. At the time, Hyneck was chairman of the Department of Astronomy at Northwestern University, and he had served for many years as the major Air Force consultant on UFOs, a position that won him little sympathy from UFO advocates. His "swamp gas" explanation of the 1966 Michigan sightings had become a national joke. Yet, his book was remarkable in two respects. First, unlike the books of Menzel and Vallee, it was based almost entirely on his own field research. The others had relied largely on secondary accounts—some of which were later shown to be fraudulent—and theoretical arguments. Second, it suggested a classification of UFO phenomena that is still widely used by UFO investigators and enthusiasts and that permitted a specialization which characterizes much contemporary UFO investigation. Hynek's books were eventually followed by other scientifically oriented works, and his Center for UFO Studies in Evanston, Illinois, carried out investigations leading to Allen Hendry's *UFO Handbook*, a comprehensive analysis of 1,307 UFO cases, of which 9 percent remained unidentified after investigation.

Doubters of the reality of UFO sightings have also been active, principally in the person of Philip J. Klass, a senior editor of *Aviation Week*, whose book (in 1968) suggested that UFO sightings were attributable to ball lightning. His next books, in 1974 and 1983, concentrated more on the possibilities of optical illusion and fraud. Energetic, inquisitive, and relentless, Klass came to provide a constant counterpoint to proponents' claims that this or that sighting was unexplainable. He was joined by James Oberg and Robert Sheaffer, both of whom later wrote their own debunking books on UFOs. Oberg, like Klass, is a science writer; Sheaffer is an engineer. All three are members of the Committee for the Scientific Investigation of Claims of the Paranormal (CSICOP), an organization set up to debunk paranormal beliefs. The critics focus on the weak points in the proponents' arguments, reinvestigate in some cases, and criticize them in the mass media. In recent years, there have been many heated exchanges between proponents and critics, in print and in person, but with little resolution of basic differences.

One interesting recent explanation for many of the so-far unexplained sightings may be that they are caused by *tectonic strain*. Michael Persinger, a psychologist, has shown that waves of UFO sightings often precede a major earth tremor. The strain that precedes earthquakes may generate fields of energy which cause luminous balls to appear in the sky (similar to "earthquake lights") and may cause people to hallucinate, the content of the hallucinations being determined by current cultural stereotypes. How many of the unexplained sightings will be explained by Persinger's theory is unknown. The correlations he has discovered suggest, however, that some UFOs may in fact be unknown natural phenomena.

The UFO investigation community, then, consists of a large number of interested amateurs, a small number of scientists, and a handful of active critics. This community has developed its own literature, its own folkways, its own jargon, and its own knowledge base. Not all UFOlogists feel that the evidence points to extraterrestrial visitation, but most believe that something very much out of the ordinary is taking place. One day, perhaps, UFOlogy will be placed alongside phrenology as a deviant and unsuccessful paradigm. Its practitioners, however, see themselves as pioneers in the recognition of the first manifestations of a nonhuman intelligence.

Is Anyone Out There Sending?

While UFOlogists are looking for extraterrestrials in the atmosphere, another group is seeking evidence in the form of electromagnetic signals from distant stars. Members of this group, the SETI (search for extraterrestrial intelligence) community, are almost all scientists who tend to think that telepathy or flying saucers are irrelevant to ETI contacts. If such contacts are to be made, some form of electromagnetic energy—radio, light, etc.—is the likely channel. This group's attention, theoretical efforts, and attempts to obtain funding thus center around the means for detecting such signals. The members of the SETI community (most visibly represented by astronomer Carl Sagan) are largely astronomers, engineers, astrophysicists, and biologists who share a common framework for discourse, although they often differ on the "realistic" estimates for the quantities which interest them. The majority, however, think that intelligent life elsewhere in the universe is very probable, if not certain.

While the notion of intelligent life on other worlds goes back to ancient times, the SETI movement is barely a quarter of a century old. Its beginning might be traced to a 1959 article, published by physicists Giuseppi Cocconi and Philip Morrison in the respected British journal *Nature,* that attracted wide attention. The next year, Cornell astronomer Frank Drake conducted OZMA, the first search for extraterrestrial radio signals. And in 1961, at a meeting convened by the National Academy of Sciences at Green Bank, West Virginia, the movement crystallized. During that meeting, one of the eleven participants, Melvin Calvin of Berkeley, was notified that he had won a Nobel Prize; this incident is indicative of the status of many members of the SETI community. Most are Ph.D.s at leading universities or prestigious research institutions. Four members of the Green Bank group have continued to work on SETI: Morrison at MIT, Drake at Cornell, Carl Sagan (also at Cornell), and engineer Bernard Oliver, who is vice-president for research at Hewlett-Packard. Another central figure in SETI research is the British physician John Billingham, now chief of the Extraterrestrial Research Division at NASA's Ames Research Center in California.

Because this group has strongly influenced intellectual discussion and research efforts, its basic assumptions are worth considering:

- First, the development of life is a natural process—merely a matter of probability, not a miracle requiring divine intervention. Thus, given the enormous number of stars likely to possess planets, the chances are extremely high that life has originated not once but many times.
- Second, the development of intelligent life is also a matter of probability and is also to be looked upon as a natural process. There are differences regarding how intelligent life is likely to develop and how closely it is likely to resemble Homo sapiens. George Gaylord Simpson, an authority on evolution, considers the chance of anything resembling man in intellectual capacity to be vanishingly small. Others, such as Carl Sagan and Frank Drake, are much more optimistic.
- Third, these civilizations are likely to carry on communications with the planets of other stars not by means of spacecraft but, rather, through the medium of electromagnetic signals. The distances between inhabited planets are simply too great to make interstellar travel common or even feasible.
- Finally, because the distances are so great, the time required even for radio signals makes two-way communication a very long process. It would take eight years for a single message and reply between Earth and the nearest star system. Thus, it is simpler and cheaper to begin by trying to detect others' signals. If these attempts succeed, then we can consider sending replies. To send messages, now, however, without knowing where to send

them, would be a waste of scarce resources. Thus, contemporary U.S. efforts focus on SETI, *searching* for ETI, rather than CETI, *communicating* with ETI.

There have been 45 searches for such signals, none successful. This lack of success is not surprising. The searches have sampled only an infinitesimal fraction of the sky and have used only a very narrow range of radio frequencies. As assumptions about what to look for and where to look change, and as the technology of search instruments steadily improves, there are likely to be many more searches.

SETI faces three obstacles: funds, competition for search frequencies with other users, and opposition within the scientific community. Funds are a serious problem for SETI. All the searches thus far have been conducted with small budgets and rather primitive data-processing equipment, and they have been able to cover only a small portion of the sky. One 1971 design, Project Cyclops, called for a system that would have cost perhaps $20 billion, but funding at this level is unlikely. The current, far more modest system, which concentrates on signal processing rather than on reception capabilities, was designed to require a start-up cost of $2 million for 1982. Initially, Congress rejected this request, but it did budget $1.5 million for 1983. These funds were to cover the first year of a five-year research and development project on a multichannel spectrum analyzer, conducted by Stanford University in cooperation with the Jet Propulsion Laboratory. The prototype system will be able to process 74,000 channels simultaneously. Five more years will then be spent on the actual search.

Another problem is that the frequencies desired by SETI investigators are also being sought for more mundane uses. For instance, the 1400-to-1727-megahertz band—thought to be an obvious "window" for interstellar communication—is also used for mobile communications, satellites, and radiosondes on weather balloons. While SETI investigators want these frequencies to be kept open for ETI searches, other interest groups have their own agendas. Furthermore, use of the radio spectrum is an international matter, and agreements have to be negotiated through the World Administrative Radio Conference, which includes 150 nations.

Still another problem is that more than one view exists in the SETI community. A strong minority position, first articulated by Michael Hart of Trinity College, Texas, holds that humans may well be the only intelligent life in the universe. Hart's view also differs from the majority in that he thinks space travel is not only a possibility but a certainty with technically advanced civilizations. If there were ETI, its presence would be manifest on earth; since there are no ETIs on earth, there is no ETI. Other "pessimists," less willing to concede the feasibility of space travel, take issue with other tenets of the majority view, such as the likelihood of other planets that could support life. The creation of intelligent life on Earth, it is suggested, required so many "lucky accidents" that one can regard it almost as a miracle. These and similar concerns were instrumental in bringing the minority together for the first "we are all alone" conference in 1979.

Although SETI has been highly publicized, the number of scientists directly involved with it is quite small. Few, if any, scientists devote all their time to it. Less than a dozen give most of their energies to projects specifically dedicated to SETI work; probably fewer than a hundred scientists are working in areas closely related to SETI. Although a majority of Americans believe that there is intelligent life elsewhere in the universe, they have devoted relatively few resources to searching for it.

Two factors account for this limited activity. First, many of the researchers active in SETI are also busy in other, more conventional research activities. Since the chances that any one individual will make a major discovery in SETI appear so slight at this time, the ambitious scientist is reluctant to divert much energy away from other activities having a higher probability of success. Second, it is very difficult to get funding for SETI research. Contrary to a prevalent misconception, SETI funding from the National Science Foundation and NASA has ranged from nonexistent to nominal. Even in its more prosperous years, the SETI unit in NASA had only one or two scientists and a small secretarial staff.

Thus, although well trained and influential, the active SETI community is relatively small. Though its funding is modest, its prestige and access to the instrumentation and talent of neighboring specialties have allowed it both to carry out some research and to publicize its activities. The impact of this group on our society testifies to its intellectual power.

Signals and Noise

In a 1977 report prepared for the U.S. House Committee on Science and Technology on the subject of ETI, one can find this interesting sentence: "Many supporters of the thesis that there are other intelligent civilizations do not support the contention that they have visited Earth—UFOs are not discussed in this report." This remark, which expresses widespread and quite strong sentiments in the SETI community, points to a striking division between UFOlogy and SETI, two human enterprises that otherwise might be seen as closely connected. Both presume highly advanced extraterrestrial civilizations as well as efforts by those civilizations to communicate with others. Obviously, such civilizations might choose more than one means of exploration or communication, and it would seem difficult a priori to determine the most likely means. Both subcultures have the problems of separating "signal" and "noise" in regard to data. Yet the prestige of these two fields of endeavor could hardly be more different, nor their participants more strongly opposed to each other's views. For all intents and purposes, these two research efforts might as well be taking place in different countries in that they appear so far apart intellectually.

This gap between the two subcultures shows up in several ways. Most conspicuous is the lack of citation of each other's work in publications. Although the work of the two groups seems complementary—after all, they are exploring different aspects of the same problem—complementarity is not acknowledged by either side. Similarly, those who attend the meetings, conventions, and briefings of one subculture seldom appear at those of the other. Finally, there are occasions when one subculture publicly attacks the other. One instance was a program in the *Cosmos* television series, narrated by Carl Sagan. In the program, Sagan's treatment of UFOlogy was largely negative; and because the series received enormous popular attention, it is to be presumed that his views were quite influential. Similarly a *Nova* program on UFOs was co-opted by the UFO critics, whose views not only were more frequently quoted but also were given more sympathetic treatment. It might be that the tendency of the news media to find an "angle" for a given program makes them tilt more toward one side or the other rather than present a balanced approach.

But there are rare occasions when the gap is bridged. Carl Sagan and Frank Drake participated in a panel on UFOs at the 1969 meetings of the American Association for the Advancement of Science. The participants included proponents, critics, and neutrals; the selection of speakers and topics was remarkably evenhanded. Sagan had also participated in the 1968 hearings before the House Committee on Science and Astronautics which dealt with UFOs, even though most of the speakers presented pro–UFO statements. Yet, despite these contacts, neither subculture appears to have had any real impact on the other. We are left, then, with a need to explain this phenomenon.

The first major cause of the separation lies in the histories of the two subcultures. UFOlogy was begun by amateurs and remains largely an amateur pursuit; this is reflected in its conferences and publications, and in the quality of the typical UFO investigation. The scientists and engineers who participate in UFOlogy have been for the most part recruited by the amateurs, and often have joined them only reluctantly. UFOlogy, furthermore, has a lunatic fringe from which it has been unable to disassociate itself. Although some UFO organizations, like NICAP, have tried to develop strict rules for what will be admitted as genuine UFO cases, the ambiguity of the data has not permitted these boundaries to be shared universally among influential UFOlogists. If one cannot definitively determine what is a UFO and what is not, cloture is very difficult. One might call UFOlogy an "incomplete paradigm." Furthermore, the realities of the lecture circuit and the media have often tempted "serious" UFOlogists to share the spotlight or the rostrum with other groups whose views they do not completely accept. The net result is that there is no well-demarcated "amateur" subculture that can be co-opted by the broad scientific community, let alone by SETI.

The SETI community, on the other hand, began and remains an "invisible college" within the scientific community, and the chosen technology—the radio telescope—simply does not permit amateur participation. (There have been amateur searches for SETI signals using ham radio equipment, but these have not influenced the public image of the SETI community.) Many of its members are scientists of considerable prestige, and even those who disagree with the majority position accept the conventions and folkways of the scientific community. The intellectual weight that can be brought to bear in favor of SETI is truly impressive. In 1982, Carl Sagan organized a petition with sixty-eight other signatories in favor of an international SETI project. The list reads like a "Who's Who" in astronomy, physics, and biology. Seven Nobel prizewinners are included. Thus, unlike the UFOlogists, whose social situation often dictates uneasy alliances, the SETI investigators have no need for compromising ties. There is, however, one tie of dependence that is important, and that is to the organs of funding. The legitimacy and the funding of SETI research is tied to its position within the scientific community, and to the prestige of its members.

This leads us to the second major reason for the gap. Not only is UFOlogy a potential contender for government funds, but any association between UFOlogy and SETI would have negative implications for SETI funding. The potential threat of such an association is that the ridicule often attaching to UFOlogy might spread to SETI research. Given the current congressional attitude that SETI research is questionable anyway, an association with UFOlogy might burden it with an unshakeable albatross. (Conversely, it would provide UFOlogy with some very prestigious defenders.)

One incident nicely illustrates the dangers to SETI of UFOlogical involvement. One participant in the 1971 debate over the supersonic transport (SST) before the U.S. House of Representatives was the distinguished meteorologist James McDonald. When McDonald gave testimony as to what he believed the effect of the SST on the atmosphere would be, his views on UFOs were brought up in order to ridicule him and discredit his testimony. He was actually laughed at during the hearing. One congressman suggested that "a man who comes here and tells me that the SST flying in the stratosphere is going to cause thousands of skin cancers has to back up this theory that there are little men flying around in the sky. I think this is very important." It does not require much imagination to anticipate how a similar association between SETI and UFOs might affect the funding chances of the former. After criticism of the current SETI microwave program appeared in *Science* magazine, one of the program's originators believed it necessary to assert that SETI "is now, and has been for decades, a scientific endeavor." Hence, whatever the private views of SETI investigators about UFOs, it is in their interest to maintain a complete dissociation in public. Because even the UFOlogists' existence is seldom acknowledged by SETI investigators, the UFOlogists, too, have reason for maintaining the separation. Why should they try to approach those who so strongly reject them?

The gap between the two subcultures, then, ostensibly an intellectual question ("the UFO evidence is just too weak"), is exacerbated by a conflict on intellectual styles and a struggle for legitimacy and funding. It is evident that some of the intellectual bases for disagreement ("the stars are too far away") between SETI investigators and UFOlogists are questioned even by some of the former (for instance, Michael Hart). Both, furthermore, can always come up with new projects: a single ETI signal or a single proved UFO case would instantly propel either activity into priority status, but to prove that there are *no* UFOs or ETIs is difficult, if not impossible. It is likely, then, that the two subcultures will continue to work in virtual isolation from each other.

We have seen in these three subcultures three distinct responses to the concept of extraterrestrial intelligence. In each case, the talent, resources, and folkways of the subculture have contributed to shaping intellectual orientations and activities. Just as it seems perfectly logical in the contactee subculture to search for quasi-religious messages from "space brothers," it seems logical to UFOlogists to interview witnesses of strange aerial objects, and logical to SETI investigators to look for electromagnetic signals from distant planetary systems. Each of these responses shows a human attempt to reach out, to understand, and to grapple with the possibility that we are not alone in the universe. Each betrays the hopes, fears, beliefs, and abilities of the people involved. In contemplating the consequences of actual contact, it might be well to try to understand the inner spirit of each of these endeavors, for they reveal a good deal about who *we* are.

Note

1. Translated by Rolfe Humphries. © 1963 Indiana University Press.

References

John Billingham, ed., *Life in the Universe*. Cambridge: MIT Press, 1981.
H. Taylor Buckner, "The Flying Saucerians: An Open Door Cult," in *Sociology and Everyday Life*, ed. Marcello Truzzi. Englewood Cliffs, N.J.: Prentice-Hall, 1968.

Ray, Fowler, *UFOs—Interplanetary Visitors*. Englewood Cliffs, N.J.: Prentice-Hall, 1974.

Curtiss G. Fuller, *Proceedings of the First International UFO Congress*, esp. part 5 (pp.261–318). New York: Warner Books, 1980.

Donald Goldsmith, *The Quest for Extraterrestrial Life*. Mill Valley, Calif.: University Science Books, 1980.

J. Allen Hynek, *The UFO Experience: A Scientific Inquiry*, New York: Ballantine, 1972.

J. Allen Hynek, and Jacques Vallee. *The Edge of Reality*, Chicago: Henry Regnery & Co., 1975.

David Jacobs, *The UFO Controversy in America*. New York: New American Library, 1975.

Carl Sagan, and I. I. Shklovskii. *Intelligent Life in the Universe*. San Francisco: Holden-Day, 1966.

Carl Sagan and Frank Drake, eds., *UFOs: A Scientific Debate*, Ithaca: Cornell University Press, 1972.

15

The Social Psychology of Space Travel

B. J. Bluth

Imagine spending six months in a moderately sized motor home going across the United States. You do not have to drive. At the same time you cannot get out and go for a little stroll by yourself. The view is matchless— 270 miles straight down to Earth and a sunrise or sunset every 90 minutes. The company, however, stays the same day in and day out.

This experience of long-term space flight is becoming routine. The Soviets have had three missions lasting between six and seven months. There is a prime crew of two men, with teams of two or three visitors (one was a woman) who come to stay for a week or ten days. The longest United States mission was the eighty-four-day Skylab, which ended in 1974. Both countries have plans for more ample stations prepared to hold larger crews. The longer the missions, however, the more difficult the interpersonal, social, and psychological situation becomes. Valery Ryumin, who has been on two six-month flights, put it this way in his diary:

> Our problem is deeply human. We must now adjust to living together, away from the rest of the world. Robinson Crusoe, after finding himself alone on an uninhabitated island solved problems by himself, accounting to himself. We have to solve ours together, taking into account the feelings of the other. Here we are totally alone. Each uttered word assumes added importance. One must bear in mind—constantly—the other's good and bad sides, anticipate his thinking, the ramifications of a wrong utterance blown out of proportion.

He added that he thought one of O'Henry's comments fits the situation quite well: "All one needs to effect a murder is lock two men into a cabin, 18 feet by 20, and keep them there for two months." There have been no murders yet, but there have been problems.

A Paradigm for Analysis

Generally derived from Talcott Parson's General Theory of Action, there are five categories that can be used to study the situations and problems that have developed in space flight.

1. *Environment*: The space environment itself, especially with its weightless condition, can affect the way a crew member works and feels. This includes such factors as weightlessness as it pertains to behavioral performance, cosmic radiation, isolation, confinement, architecture, technology, tasks, ground control, and the outside world of service, family, friends, other groups, and events. This category also includes the design of equipment and the concern for human-machine interface.
2. *Physiology*: Living in the space environment, a crew member experiences significant bodily changes that affect many of the sensory systems as well as physical and hence mental tone. This includes the impact upon the skeletal, muscular, and cardiovascular systems. Space sickness (though it is usually limited to three to seven days) is a clear example of impaired effectiveness created in the space environment. Other factors that may affect human mental tone by means of physiological changes are related to types or availability of various foods, or ingredients of the life-support system.
3. *Personality Systems*: Individual predispositions of cognition, attitude, motivation, values, and thinking patterns can contribute to error and poor adaptation. Because an individual changes in both predictable and unpredictable ways, a selection procedure based on personality tests is a useful but far from flawless way of choosing astronauts.
4. *Social Systems*: The design or organization of systems of authority, roles, schedules, communication patterns, decision making, leisure activities, jobs, and informal interaction is related to the way in which a crew performs its duties. This category includes systems of rewards and career advancement, manipulation within and interchange with the environment, social control, and socialization processes.
5. *Culture Systems*: Ideas, values, beliefs, assumptions about what is good, bad, important, and real are ongoing background factors in crew activity and ground-crew interchange. These background factors can have an important effect upon individual and group performance. Expectations about the varied demands of a space mission can also shape attitudes and behaviors of crews.

Space Flight Experience

Both the U.S. and Soviet space achievements have been impressive, amply demonstrating the ability of people to live and work successfully in the unique environment of space. However, both programs have also had their share of problems that can be traced to social and psychological factors.

Soviet Experiences

Two missions of approximately six months duration and one of almost seven months have been flown by crews of two Soviet cosmonauts in the Salyut 6 and 7 space stations. Prior to each mission, extensive tests and activities are undertaken by the Group for Psychological Support to ensure the compatibility of the crew members and provide training for the flight. Crews undergo a rigorous program designed to develop self-confidence and emergency survival skills. During each mission the Group for Psychological Support constantly monitors the crews to detect signs of stress, and is responsible for developing measures to compensate. In spite of all these precautions, General G. Beregovoi, chief of crew training, reports that crew members develop signs of interpersonal hostility about 30 days into the mission, and this situation is a matter of concern for flight safety and mission effectiveness. This problem is confirmed by cosmonauts. Cosmonaut Sevast'yanov commented that "joint existence cannot be serene. We had disagreements in flight. . . . The disagreement did not reach a scandal, inasmuch as there was no 'platform' [it would be very hard to "punch" someone in weightlessness], it was simply fatigue, and frequently simple inattention which could cause the argument." Cosmonaut Lenov remarks that "the effect of psychological compatibility arises after approximately one month of staying under conditions of group isolation. The longer the flight, the more strongly the given effect appears." He concludes that in spite of preparations and training, "it is impossible to ensure good psychological compatibility on a long flight." And there are rumors that cosmonaut Ryumin wanted to return to Earth a couple of months into his second long flight. Soviet scientists have speculated that the cause of this problem comes from exhaustion of the nervous system, the biological toll of weightlessness, informational overload, and constant publicity. Other studies of Soviet systems suggest that social and culture systems variables are important contributing factors, and as such indicate that remedies are possible.

The hostility that has appeared in Soviet missions is not confined to the flight crews. There also have been arguments and conflict between the crews and the mission control staff on the ground. One cosmonaut commented that sometimes they were happy in "autonomous flight without communications from Earth. To be alone without constant nitpicking from the Earth was sometimes necessary." The crews were reported to have "masked" their feelings and reactions, withheld information, and on one occasion were rumored to have turned off all communications with the ground for two days.

The crews show mood swings, increases in tension, and difficulty sleeping. They are apparently reluctant to use the games provided by the Group for Psychological Support for them to relax, and often use their leisure time to work. "Rest and work in space has a different meaning," according to cosmonaut Kovalenok. In space you want to "load yourself with work so the time will go faster. Otherwise, you feel that the time slows down," and then you feel the loneliness, or you start thinking of aches, pains, sinus congestion, or your general physical condition. Lack of work to do is related to psychological conditions, which he says are "particularly severe under zero-gravity conditions." In space, a person is in a state of "constant alertness."

An example of a case of exuberance that nearly resulted in tragedy for the Soviet program occurred when two cosmonauts, the rookie Romenenko and the veteran Grechko, arrived on Salyut 6. There was some fear that the automated Progress supply ships might not be able to dock at the station, which would have forced the crew to return to Earth, so Grechko made an early space walk to check the docking port. Romenenko, suited up, was in the airlock monitoring Grechko's life-support system. He had heard about how beautiful space can be—stars like jewels sparkling on black velvet, and colors so beautiful that no photograph could imitate them—and the story goes that he excitedly decided to take a little peek, just put his head out the hatch for a look. Because he had not secured his safety tethers, when he pulled himself up his momentum in weightlessness took over and he found himself flailing around, gradually floating away from the spacecraft. Fortunately, Grechko saw him and grabbed his foot, saving him from death in deep space. It was more than a close call.

Soviet cosmonauts, like their U.S. counterparts, are trained to follow checklists and comply with safety requirements without fail. Such a lapse of judgment was a serious mistake, transcending Romenenko's small adventure, for it raises questions about the way some people may react in space.

V. Vereschetin, a deputy chairman of the Intercosmos Council, told a Moscow news conference that "cosmonauts' efficiency declined during long flights." Cosmonaut Ryumin (who was on both long flights) believes a human being could stay longer but was quoted in *Isvestia*, 28 March 1981, as saying that "after three to four months of work on orbit, exhaustion sets in. And if there is not a special need for some sort of very important task, there is no particular sense to push a cosmonaut's organism to its endurance."

At a recent conference in Bonn, West Germany, Academician Gazenko reported that "psychological and every-day life discomfort . . . seems to be inevitable for space operations for the near future," and in this context the "duration of space missions may be dependent on nervous-emotional stability of crew members, their psychological compatibility, level of motivation, pattern and stability of various habits, and other factors that can hardly be identified.'' The question is, why do cosmonauts become fatigued, what affects their stability, and which factors can be identified and changed?

U.S. Experiences

Accidents, near-accidents, and problems that can be traced to the complexity of social, task, and personal factors have also occurred in the U.S. space program. One serious incident took place on the Apollo-Soyuz Test Project as the command module reentered the atmosphere. There were several cockpit errors on this mission, including forgetting to set two reentry switches that prevent automatic devices from deploying the chutes and from dumping excess fuel and oxidizer from the attitude-control rocket engines. Commander Tom Stafford apparently was distracted and neglected to call for the setting at the proper time; rookie astronaut Vance Brand also missed them. As the capsule reentered the atmosphere nitrogen oxide gas filled the cockpit. Brand lost consciousness, and it took five minutes for Donald Slayton and ''Deke'' Stafford to unstow the oxygen masks. The effect of such a toxic mixture on the lungs could have been fatal.

In evaluating the reasons for this error and the others that occurred, astronaut Walt Cunningham suggested that Stafford and Slayton had not trained as much as they should have for the mission and that ''the crew did less training *together* than usual, even for those phases of the mission which require close coordination.'' He cited a heavy social schedule, emphasis on public relations, and the fact that much of the training load was placed on rookie Brand's shoulders. There was a significant status difference between Slayton and Stafford on the one hand and Brand on the other. Slayton and Stafford were veterans of the space program with many years of experience and high positions in NASA; Brand was young and new. The near-accident was caused not by equipment failure but by human error conditioned by social system factors in conjunction with personality and cultural factors: the value placed on social and political priorities superseded the need for training, and the influence of status may have inhibited the communication process.

Scheduling and time pressure are cited as some of the major causes of error on the Skylab IV mission, which lasted 84 days in 1973–74. Because it was to be the last mission for Skylab, many new experiments were added to the flight plan. The crew had not trained for these experiments, and thus took longer to set up and carry out the experiments than had been expected. There were many unanticipated difficulties, yet the ground pressed the crew on so as to take advantage of the precious time in orbit. The result was that the harried crew was plagued with errors, mistakes, and lost data. The problem got so bad that Commander Jerry Carr asked Mission

Control to have the ground staff completely rework the program. Crew members were given more discretion in scheduling, a shopping list of some experiments, and more time to themselves. Performance improved considerably.

Contributing Factors

Early studies of the social and psychological factors involved in long-term space flight started from the assumption that the systems, errors, and accidents that occurred in space or related environments were primarily related to the biological ramifications of weightlessness or the phenomena of isolation and confinement. The symptoms that were found in either Soviet or U.S. space flight included insomnia, irritability, mood fluctuation, fatigue, hostility, vacillating motivation, depression, and stress—symptoms quite similar to those found in isolated and confined environments. The symptoms also have much in common with those that develop in environments that are not isolated or confined but are associated with the development of stress. Although some of the space flight symptoms can be traced to the effects of weightlessness on the body, and vary in intensity as the body adjusts, there are other systems that can be related to the structural aspects of stress in general, and as such are subject to alteration.

Symptoms of Isolated and Confined Environments. Environments studied by NASA in the 1960s which were considered to be isolated and confined included the Antarctic, submarines, oceanographic research vessels, undersea habitats, simulations, and fallout shelters. NASA itself sponsored many simulations as well as undersea expeditions to research the features of isolation and confinement. The symptoms that can be summarized from that literature are as follows:

- Boredom
- Irritability
- Depression
- Anxiety
- Mood fluctuation
- Fatigue
- Hostility
- Social withdrawal
- Vacillating motivation
- Tension
- Sleep disorders
- Lowered performance

Symptoms of Stress. The general stress literature showed the following symptoms:

- Boredom
- Irritability
- Depression
- Anxiety
- Excessive emotion
- Anger
- Tension
- Sleep disorders
- Lack of concentration
- Lowered efficiency
- Defensiveness
- Fear

This research traces the causes of the symptoms to myriad sources, many of which can be related to space flight situations. Some are: conflicting definitions of a situation, mismatched work, leadership systems, scheduling, expectations, group size, reduced roles, reduced sensory input, architectural arrangements, group composition, training, communication systems, physiological factors, crowding, loneliness, career development, job design, time pressures, and long hours.

Both the NASA research and general stress studies suggest that many of the symptoms that develop on long duration flights may not be unique to the space environment, and in fact may be alleviated by training and careful design of social and environmental features. A few examples from this vast literature follow.

Physiological Factors

Weightlessness brings about major changes in the human body, and there should be no discounting the potential impact those changes can have on the psychological readiness of a space station crew. However, Earth-bound studies have shown that people can mistake socially or psychologically created tension as a biological problem. For example, fatigue and sleep disorders are often a symptom of social or psychological stress. Continued disruption of the sleep cycle increases fatigue while increasing general tension, and thus induces more strain in the situation—a vicious circle.

A recent report from NASA also shows that there are clear psychological effects from alterations in circadian rhythms and the rapid changes from light to dark that occur on orbit. Symptoms included depression, hopelessness, boredom, irritability, negative moods and withdrawal, fatigue, decreased performance, insomnia, anxiety, and gastrointestinal and other physical symptoms.

Social and Psychological Conditions

Including the experiences to date, but also considering some of the new dimensions that will turn up in larger, more mixed crews, other social and psychological factors appear important to the quality of group relationships.

Conflicting Functions. Normally, a person's involvement in a group represents but one facet of his or her life. There are many groups, many roles, many varying social demands, and thus many outlets for interest and sources of support. The work group where one is constantly evaluated is distinct from the family or friendship group where one believes that one is accepted and wanted for oneself. Ideally, one need not be on guard, or prove oneself, or put forth one's best effort. In the small isolated and confined space crews, where work and informal life are combined, all the aspects of daily life are carried out in the presence of the same group of people who are always on hand. One is never alone. The "boss" is nearby. Such a situation automatically places a strain on the members because the group must fill two different and incompatible functions: work and friendship—evaluation and nonevaluative support. In a space station, additional strain derives from potential danger; survival mandates interdependence. Research shows that such a situation can bring on fatigue, headaches, poor communication, and problems with authority. To cope with this problem in its small crews of two, the Soviets have almost completely abandoned formal hierarchical roles between crew members. The Americans have clear hierarchies in place for the Space Shuttle missions, though it is questionable how salient the hierarchies now are since these crews have worked together for so many years and seldom use formal role structures in their interactions.

Group Size. There is evidence that groups with an even number of members differ from groups with an odd number of members. Even-numbered groups disagree more than odd-numbered ones and suffer more deadlocks. Groups with an even number may split into equal factions, which cannot happen in an odd-numbered group. Groups of five, for instance, are shown to rate high in member satisfaction. Dissenters have some group support. As the group gets larger, there is more stability but the lines of communication tend to become focused on the leader, with some members becoming more passive. When the groups are total systems, with work and primary relationships being coextensive, the occurrence of deadlocks and disagreements are more frequent and severe in their consequences. However, there is a greater tendency to soften rules or circumvent the command structure to overcome these difficulties. Total system situations can become a source of strain and error-producing conditions. Some structural generative elements of stress can be avoided if group memberships are in the odd numbers. (Five, nine and eleven are especially workable sizes.) This seems to have been confirmed in both the Soviet and U.S. experiences. The Soviets, with crews of two, seem to have had more interpersonal problems than the U.S. crews of three, a fact which was predicted in the early 1960s by NASA researchers.

Group Composition and Rotation. Newcomers in small, cohesive groups often complain of being shut out. A basic trait of group membership is the feeling of togetherness, belonging, and a sense of "us" as opposed to "them." The identity of the group becomes defined by reason of its distinction from other groups, and in a desire to preserve its identity, the group often forgets internal disputes while confronting "outsiders." In fact, this is often used as a method of provoking internal group cohesion and loyalty when the group itself may be showing signs of instability. Some discomfort has been reported by the non-Soviet cosmonauts who have been visitors to the Salyut station in the Intercosmos Program. The Czech cosmonaut Remek reported feeling uncomfortable and left out on occasion. There were misunderstandings, hurt feelings, and a genuine concern about the problems of language and communication. Thus the prospect of rotating crews on a station, half leaving and half staying, may create two separate and possibly antagonistic groups. If schedules further tend to isolate them from each other, while still living and working in small quarters, the potential for distrust, suspicion, and hostility is increased. There may be many unexpected problems if the crews are mixed cultural subgroups or come from multinational backgrounds.

Group Organization. Research in the area of organizational development has shown that organizational structure should vary with the job. Jobs involving research require a more open-ended, loosely structured system; jobs that require on-site problem solving need flexibility; and jobs that are precisely defined and broken down into small divisions of labor demand a high level of structure. The formal aspect of an organization, which is the fit between its task and its formal practices, needs to match the climate of the organization, which refers to the subjective perceptions and orientations people have about the organizational setting. These are compared on predictable and unpredictable dimensions. Thus predictable tasks call for a predictable climate, and unpredictable tasks call for more fluidity and flexibility in the climate. On Skylab, however, many unpredictable jobs were scheduled as though they were highly predictable, creating stress for the crew and tension between the crew and ground control personnel who were responsible for the planning.

Numerous studies have also shown that mental health problems, stress, and alcoholism are related to a lack of participation in decision making, a lack of control of the work, or a lack of autonomy. Stimulating some internal regulation of the system by the group, using discretionary rather than prescribed work roles, increasing variety, and decreasing the bureaucratic mode, while keeping in mind that the primary function of these groups is to work, have resulted in significant increases in effectiveness and a decrease in stress-related factors. When Skylab crews were given more control over their schedules, their productivity improved significantly. A look at some of the problems that developed on Skylab suggest that a different design of the crew's work and responsibilities might have alleviated some of the difficulties.

There were also disagreements between the scientist astronauts and the test pilot astronauts during the Apollo programs that resulted in lost, damaged, or eliminated experiments. Such subculture differences could also be a problem on long-term missions if the organizational and decision systems are not designed to recognize the varying assumptions, values, and requirements of the participant groups.

Technical and Architectural Design

Architecture. Research in this area shows that the design of work spaces and living quarters can have an important effect on mental state. Astronauts and cosmonauts have commented on color (they do not like what is there), windows (there are never enough), noise (there is too much), smells (there are not enough good ones), flexibility in arrangements (things cannot be changed), restraints (there are not enough places to put things, and things do not stay put), traffic flow, organization of public and private room (there is no place to be "alone"), and the inclusion of "nice" touches such as fish and plants.

Technical Design. Studies show that the sociotechnical design of work stations and computer terminals should fit the kinds of mental models needed to operate the facility efficiently. These studies are proceeding in the design of aircraft cockpits, nuclear power plants, and general areas. The work shows a need to identify, *before* the equipment is built and while it is being designed, the mental models needed to operate it under both routine and emergency conditions. Analysis of Space Shuttle and Skylab equipment suggests that some of the features are troublesome and conducive to human error or difficulty of use in weightlessness, and in need of redesign.

Current U.S. and Soviet Activity

Although there is ample evidence of the importance of social and psychological factors in long-duration space flight, the United States and the Soviet Union have completely different approaches to the problem.

U.S. Activity

In the early days of the space program there was considerable concern about the possible psychological effects of space, and thus numerous tests, studies, and reports were made. Space did not turn out to be as mentally debilitating as was feared, and as the astronauts garnered the power to fly spacecraft instead of ride them, they also edged out of the psychologists and their programs. They seem to have had good reason for these actions, for reportedly the psychologists were not very sensitive about the feelings and attitudes of their subjects. Suggestions were often seen by the crews as too theoretical and lacking in practical application. The result was that there were *no* systematic psychological studies made of the Apollo or Skylab crews, or of conditions on board the spacecraft that related to mental state, error, or the like.

Psychological studies dropped off dramatically but did not stop. NASA continued to sponsor work at Johns Hopkins on the kinds of conditions that developed in confined microsocieties of two or three members, a management study of a spacelab simulation, evaluations of undersea conditions and their effect on mental and interpersonal activities, and some research dealing with situations on oceanographic research ships and male/female crews on supertankers, among other subjects. The results, however, did not filter into either the operations design process at NASA, and hence have had little impact on flight programs.

Now that NASA is seriously planning for a space station within the next decade, there are people in the agency who are concerning themselves with the social and psychological problems involved in long-duration missions. Some small studies being done in-house and by industry on social and psychological factors include the command and control of varying numbers of mixed crews, work and rest schedules, the influence of diet on astronaut mood and performance, and the impact of small quarters and remote locations. The researchers are also reviewing past work in the behavioral sciences to determine where there are gaps, and are identifying steps that need to be taken in these areas.

Soviet Activity

The Soviets have been deeply involved in the psychological aspects of space flight since the inception of their program, and their approach is a vigorous one. They state quite clearly that long-term space flight is highly dependent on the psychological state of the crew, and they are taking every measure they can to ensure that it is good. They engage in research, testing, training, in-flight observation and troubleshooting, and postflight evaluations. The Group for Psychological Support does much of this work.

It is constantly involved in sophisticated voice-stress analysis while the crew is in space. When high levels of stress are indicated, or are growing, it arranges interactive television meetings with family, friends, famous people, and the scientists involved in the experiments the crew is conducting. It has arranged for surprise packages to be delivered in the Progress resupply ship in addition to special foods, fresh fruits, letters, video movies and special programs, gifts, and even a guitar.

One purpose of the Intercosmos Program, which has brought cosmonauts from many non-Soviet countries to the space station, is to break the boredom, monotony, and loneliness of crews on very long missions, in addition to helping them meet technical goals. Parties are celebrated with ritual salt and bread, the customary foods of the visiting country, and so on. The Salyut 7 spacecraft also included changes in colors and design to make life aboard more pleasant. All the cosmonauts go through a rigorous training program intended to develop self-reliance, self-control, self-confidence, and the ability to keep calm and alert in an emergency.

In spite of all these efforts, the Soviets still report many of the symptoms and problems mentioned above. They state publicly that they have not developed adequate methods of establishing compatibility among crew members and that much of the adaptation is still done on a common sense level. They also are showing serious interest in the effects of weightlessness on moods, attention, and motivation.

The overall Soviet attitude seems to be one of providing the crew with a variety of possible alternatives in terms of skills, knowledge, and psychological training to cope with long-term space sojourns. The Soviets assume that human beings are not limitless in their capacity to adapt; success demands a good understanding of strengths, weaknesses, and boundaries so that people and their environments can work synergistically. To leave the human element, one of the most important factors in the "loop," to chance or common sense seems to them to be hardly a scientific approach if mission safety and success are the goals.

It is very hard to evaluate the Soviet material, even when it is available. The Soviet's have a different approach to data than Westerners do, and thus do not treat or include information that might be thought necessary from our point of view. Their research which is known in the West appears to be heavily oriented to the biological and psychological factors, with little emphasis on social systems or some of the other factors noted earlier. In psychological training, testing, and study, cosmonauts seem to be treated from the psychological perspective, not as members of a vast system of layers of social strata and group dynamics. Of course this may not be the case. The Soviets claim quite vigorously that they were for many reasons not planning to send more women into space for quite a long time—but then Svetlana Savitskaya was placed aboard the Salyut 7 in August of 1982.

Conclusion

Spaceflight experience, especially by the Soviets on very long missions, and data from isolated, confined, and stressed populations show that the longer a flight, the more critical social and psychological factors become in flight safety and mission effectiveness. All of which also show the importance of the human factor in space. As space flights continue, the involvement of practical behavioral sciences will no doubt increase, giving rise to unprecedented opportunities for applications and research. Space stations are natural laboratories. Situations are real, are small, and it is possible to manage the amount of information involved. Crews may be willing to try alternative social systems, and thus we can learn much about basic fundamental social processes without the limitless known and unknown variables that interfere with Earth-based research. The frontier of space is also the frontier of the social sciences.

References

B. J. Bluth. "Consciousness Alteration in Space." In *Space Manufacturing 3: Proceedings of the Fourth AIAA/Princeton Conference, May 14–17, 1979*, edited by Jerry Grey and Christine Krop, 525–32. New York: AIAA, 1979.

B. J. Bluth. "The Psychology and Safety of Weightlessness." Paper presented at the 15th Symposium on Space Rescue and Safety. International Astronautical Federation Congress, Paris, September 1982.

B. J. Bluth and S.R. McNeal. *Update on Space*, vol. 1. Los Angeles: NBS, 1981.

Walter Cunningham. *The All American Boys*. New York: Macmillan Co., 1977.

James Oberg. *Red Star in Orbit*. New York: Random House, 1981.

Sherman P. Vinograd. *Studies of Social Group Dynamics under Isolated Conditions*. NASA CR-2496. Washington, D.C.: December, 1974.

Tom Wolfe. *The Right Stuff*. New York: Farrar, Straus & Giroux. 1979.

16

Must There Be Space "Colonies"?
A Jurisprudential Drift to Historicism

George S. Robinson

Space Law and the Lulling Effect of the "Peace Image"

For years, a handful of well-meaning lawyers, diplomats, and assorted bureaucrats worldwide have been nudging a body of principles, rules, and general statements of international accord into an amorphous body of wishful thinking called space law. It is *wishful*, or *precatory*, as we say in the legal profession, because events show us that few countries have any intention of being bound by elevated legal principles of peaceful exploration and use of space, particularly if those principles should prove to be obstacles or embarrassments to practical development of national space capabilities.

The Outer Space Treaty of 1967, dubbed "The Mother Treaty" by lawyers (and more properly but laboriously called the United Nations "Treaty on Principles Governing the Activities of States in the Exploration and Use of Outer Space Including the Moon and Other Celestial Bodies), serves as the nucleus of most international agreements relating to space activities. Among the principles encompassed by the treaty are:

- Space exploration shall be conducted for the benefit of all countries, and shall be the province of all humankind.
- Outer space and celestial bodies cannot be claimed by any country for itself.
- Space research is to be carried out in the interest of furthering international cooperation, understanding, and peace everywhere.
- Outer space may not be used for the placing of nuclear weapons or other weapons of mass destruction, nor shall there be any military bases, installations or fortifications, maneuvers, or weapons testing in outer space.

- Astronauts shall be considered envoys of humankind and shall be given assistance and protection in their endeavors.
- States, governments, and international organizations shall have certain liabilities for activities and accidents arising from space exploration.
- Efforts will be made to avoid contaminating celestial bodies or harming Earth's environment as a result of the introduction of extraterrestrial matter.
- The signatory states must consult with one another when there is reason to believe a planned experiment would cause harmful interference with other activities or interests of the signatories. They must consider requests for visual observations by member states on the basis of equality regarding various activities, the purpose being to encourage international cooperation in space research.

Flowing from these commendable principles and good intentions are several other international agreements that have been negotiated and signed into effect, among them the Agreement on the Rescue of Astronauts, the Return of Astronauts and the Return of Objects Launched into Outer Space; the Convention on International Liability for Damage Caused by Space Objects (fairly recently put to the test with the descent of a Soviet nuclear-powered satellite into Canadian territory and the unplanned orbital decay of our own Skylab IV); the Convention on Registration of Objects Launched into Outer Space; and the recently negotiated treaty for exploiting the resources on and beneath the moon's surface.

Lawyers have given very little attention to the special needs and interests of astronauts and others living in a synthetic life-support and often hostile environment. Almost no consideration has been given to the special legal regimes for inhabitants of space manufacturing facilities and other forms of permanent or long-duration space communities that are sure to come. Perhaps the reason lies in the embryonic nature of economic theory that will attach to establishment of space communities. At the outset, it is the economic reality of such a large and costly venture, as well as pressing military requirements in space, that will dictate most of the lifestyles and, also, the social and political structures of, say, a permanently manned space manufacturing facility or orbiting military station.

In the minds of most people, the idea of Earth-orbiting space communities is still a very distant possibility with equally as distant problems to be faced. That the United States already has established several long-duration space communities in the Skylab program and the ongoing Space Shuttle flights, and the Soviets a number of others—including the first cosmonaut crews of mixed sex and mixed nationalities—is still difficult for many people to grasp. So far, these events are really perceived as only technological accomplishments. But there is a pressing immediacy to focusing constructive attention on the social requirements of people living in space. The Soviet Union has launched its second long-duration space station with a view to extending the habitat system permanently, just as soon as the technology permits; the European Space Agency has built its manned space laboratory; and the next stated goal of the U.S. National Aeronautics and Space Administration is a permanently manned space base.

The idea of very large communities numbering individuals in the thousands and hundreds of thousands still seems to belong to the strange realm of science fiction, but that is not true. There must be small steps in this direction, and the idea is being developed in comprehensive and detailed fashion, both by the United States and the Soviet Union. Unfortunately, it is being developed within the political deceptiveness of a peaceful space activities image.

Most people anticipating and preparing for the circumstances surrounding development of large, manned space facilities and habitats have been guided, perhaps even constrained, by the "peaceful-use-of-outer-space" ambience fostered and cultivated in the Outer Space Treaty. In attempts to elevate space exploration and occupation to a higher level of human dignity and integrity than that characterizing the evolutionary history of civilizations on Earth, some economists, political scientists, engineers, cultural anthropologists, lawyers, and assorted social and natural scientists have conceptualized space facilities and societies as sanctuaries from the ever-grinding conflicts of civilizations on Earth.

Of course, we must continue applying ourselves toward that end. But we must not ignore the historical lessons of imperialism, militarism, and economic colonialism, and the strong possibility that they will also characterize human behavior and culture evolving in outer space. We must not ignore the glaringly antithetical lessons of history that emphasize the ultimate *violence* of colonialism and militarism as the primary tools of cultural and economic imperialism. Put differently, can we afford to ignore these lessons? It may well be that the films *Star Wars* and *The Empire Strikes Back* are not only entertaining but may also represent the simple declaration that the nature of humankind will not undergo some elevated evolutionary change while transitioning from Earth to near and deep space.

Article 4 of the Outer Space Treaty provides indirectly for military use of space as well as peaceful participation in its exploration by asserting in part that

> State Parties to the Treaty undertake not to place in orbit around the Earth any objects carrying nuclear weapons or any other kinds of weapons of mass destruction, install such weapons on celestial bodies, or station such weapons in outer space in any other manner. The testing of any type of weapons . . . on celestial bodies shall be forbidden.

Out of Article 4 emerges the most difficult issue of deciding what is acceptable military use of, or involvement in, outer space activities. Unfortunately, but perhaps realistically, the answers to this issue seem to be formulated only by the so-called superpowers in their negotiations of such strategic agreements as the ABM Treaty of 1972, which limits the capabilities and uses of anti-ballistic missiles; the 1972 Interim SALT Agreement, purporting to limit the use of certain strategic weapons; and the 1963 Limited Nuclear Test Ban Treaty, which attempted to force nuclear testing into underground sites.

All of these agreements encompass military use of outer space covered at least in part by the principles, spirit, and intent of the 1967 multinational Outer Space Treaty—but without the treaty's positive terms of reference for peaceful exploration and exploitation. In short, the vast majority of nations that signed the Outer Space Treaty but that have no present space capabilities of their own are being disfranchised from their rightful participation in determining precisely what is acceptable military use of space. That determination is being accomplished through the negotiation of a few agreements by those few military superpowers that have the necessary technology, or access to it, for placing military weapons and intelligence gathering devices in space.

Because the United States, the Soviet Union, and a handful of other nations are negotiating their own definitions of acceptable military and economic uses of space, we should look more realistically to the military–political historian for our guidelines in planning economic, political, and legal regimes for space habitats and the sophisticated exploitation of space resources. In fact, we might best look at those perspectives and lessons of colonialism reflected in known legal regimes. As so aptly described by Lawrence M. Friedman in *A History of American Law* (New York: Simon & Schuster, 1973), "The law is a mirror held up against life. It is order; it is justice; it is also fear, insecurity, and emptiness; it is whatever results from the scheming, plotting and striving of people and groups, with and against each other. . . ." Friedman perceives a full history of U.S. law as "nothing more or less than a full history of American life. The future of one is the future of the other." What better history to use as an indicator of the unfolding human occupation of outer space than the history of New World colonialism and its legal institutions?

American Colonies and Space Habitats: The Legal Twain Shall Meet

Most historians seem to be insensitive to the significance of legal history and its incredible wellsprings reflecting values of a society at any given time. Lawyers and jurists continue to write their own histories of legal principles and regimes, and their theses are little more than esoteric battlefields of concise theories—seemingly unrelated and fictionalized—won or lost in the judges' chambers, or bargained to a settlement over beer and pretzels.

Students of Anglo-American law are taught that tradition, or precedence, is the notochord of common law, the body of law established over time through case-by-case decisions, and not by a legislated, all-encompassing "code" of law. Most Americans still believe that tradition is strong in the various legal systems of the United States, and that the traditions of English law formed the legal regimes of most colonies. That is not altogether true.

Certainly, various aspects of contemporary law trace their roots back rather far into English history, the jury system, trusts, mortgage concepts, and the still-current use of English covenants "running with the land," those provisions relating to the quiet and unhampered enjoyment of title whenever land is purchased or otherwise transferred. But in practice, law is very organic, constantly growing, molting, and undergoing various types of metamorphoses to accommodate (1) constantly changing cultural values based upon the scientific and technological fruits of human intellect; and (2) the frequent turnover of political regimes and underlying philosophies. Ponderous political and legal frameworks cannot accommodate the ruthlessly inexorable demand for changes in legal systems birthed by evolution and revolution. Tradition cannot, for example, accommodate the need for no-fault insurance and divorce laws, tax laws, health laws, and criminal rehabilitation laws. And this seems certainly true for the economic, political, and social behavior requirements for humankind functioning in a totally synthetic and alien life-support environment of a space habitat. Old traditions and values stay alive in a new environment only if they serve a purpose or do not interfere unacceptably with the legal mainstreams of a society.

Anglo-American common law develops slowly. It is impotent to meet fast-changing requirements for social stability. In fact, at the outset of space exploration in the late 1950s, extensive consideration was given to establishing a "Space Code" as the only way to control the then-foreseeable fast-changing nature of space activities. But after much struggling and agonizing over this jurisprudential concept—with which the Soviets, French, Germans, Spanish, and Italians were most familiar—a hiatus was reached where the lack of practical problems and experiences in space capabilities and activities forced legal minds to establish the initial common law of space. The first principle of such law characterized sovereign airspace of subjacent states as having definite upward limits. This principle was based on the fact that no nation objected to the Soviet launching of the first Sputnik. Later, the common law of space became embodied in codified international law, i.e. formal treaties.

Much in the same manner as we are currently discovering in our progressively sophisticated manned space programs, the daily requirements of a few soldiers of fortune, merchants, and persecuted religious zealots barely surviving in settlements along the eastern shores of a continental North American wilderness were substantially different from those of highly commercialized and industrialized countries several thousand miles away. The legal needs were worlds apart also. But because the small settlements had no indigenous cultural pasts of their own, it was much easier to import elements of legal and economic theory that were familiar and required no, or little, translation for immediate use.

In many respects, the same type of attitude, i.e. "importation of the familiar," understandably characterized the initial habitats or "colonies" in Earth orbit and on the lunar surface. Unfortunately, it could also be the approach followed for long-duration or permanent habitation of space by humankind, with little consideration given to the unique behavior patterns and social requirements of a confined and alien synthetic life-support environment.

Despite our United Nations efforts to the contrary, we are beginning to see strong indications of legal nationalisms and parochialisms already occurring in the conduct of space activities, e.g. the assertion of sovereign rights over geostationary orbits and other highly prized locations in space by small Third World countries with no present space exploitation capabilities. In fact, we are very likely to continue seeing political pressure for limited mercantile and social laws unilaterally established by a launching state; or perhaps even a politico-economic consortium underwriting a manned space venture much in the same manner as an industrial "company town," in which the underwriters and not governments are largely responsible for the social behavior and welfare of the employee inhabitants.

Empirical Nature of Colonial Law

English law transferred to the North American colonies was pluralistic. For the most part, it consisted of common law and formal court proceedings for the aristocracy, as well as an expedient customary law that was developed to give order to the routine business of family life and local commerce, particularly in frontier settlements. The formal system consisted principally of courts of law, courts of chancery, and the mercantile courts, the last of which embodied the law merchant, an international set of rules sensitive to business transactions and deriving "from the general customs and law-sense of European traders."

The comparison of these formal and customary laws and procedures with what we can anticipate in legal systems of space communities, might be the following:

1. Common law emanating from king and parliament would be the equivalent of multilateral and bilateral space treaties produced by public international

organizations like the United Nations, North Atlantic Treaty Organization, and perhaps even the formal arrangements of the Warsaw Pact countries.

2. Equity law, deriving from the Courts of Chancery, dealt, of course, with affairs of people and not with direct jurisdiction over "things." Equity law was more than the everyday backwoods law relating to relationships among people whereby the prevailing spirit of fairness and justness regarding those relationships are determined by peer groups in a comparatively structured court system. In North American colonies, the common law and equity law were in constant combat for dominance, and the formal common law survived only through deep and extensive compromise with the demands of equity and the growing laws indigenous and responsive to the unique requirements of colonial settlements. Equity law born of the very special circumstances of space communities could well demand deep compromises in the application of space treaties prefabricated on Earth and shipped off to space habitats.

3. The mercantile courts applied those laws peculiarly responsive to the international trading needs of merchants. They were courts that spoke the strange language of international finance and trade; and they disseminated quick justice to encourage and expedite trade. Just how the mercantile courts will equate with the institutionalized methods of trade among space communities, and between those communities and Earth institutions, is speculative. But some reasonable speculations can be entertained based upon the historical patterns of those courts.

The physical and economic survival requirements of small settlements tenaciously gripping a hostile coastline on an unknown continent several thousand miles from European civilization were substantially different from those which prevailed by the time independence from England was declared. The law that was applied to the burgeoning commercial, political, and other relatively sophisticated activities in the early 1770s tended to be imported. In the isolated backwoods trading posts and settlements of that period things were happening too fast by then for the responsive but time-consuming evolution of law. Instead, the source of importation was from those countries with the same language and traditions of the colonists. Familiarity and shared language were the primary selective factors.

Again, as observed by Friedman in *The History of American Law*, the colonies just before independence combined local and traditional legal principles to order their social and commercial routines. "Both colonial law and the law of the United States," he asserts, was spawned of "uniformity and diversity in constant tension over time." The fabric of "colonial laws" was of a richness in diversity because the climatic, geographic, political, and theological conditions, and structures of the colonies were diverse in the extreme.

And on top of this, courts were "parchment"—that is, what the judicial structure theoretically ought to have been, not what they were, or would be, in fact. These circumstances and practices are very similar to the present-day activities of lawyers, statesmen, and politicians who are busily establishing legal regimes and structures for outer space exploitation and occupation, regimes that are not now, and very likely never will be, responsive to the actual circumstances of humankind's permanent movement into space.

Both before and immediately after independence, much of the formal law was established, not to regulate routine relationships and commercial activity but to fit the perceived political needs and policies of England and the colonies. Our space law has much the same foundation, only it has evolved in large part, until recently, from wishful anticipation and not from a hard-nosed factual review of economic, political, and military circumstances relating to space capabilities.

From this broad characterization of colonial legal regimes as a reflection of the fears, insecurities, schemings, plottings, and strivings of "strangers in a strange land," hostile and precarious but with unlimited development potential, we can look at the establishment of space communities through the eyes of a legal historian and perhaps determine measurably how these communities will be established in fact, as opposed to how we think they ought to be.

Maverick Commercial Offspring, or Legally Ordered "Colonies"?

Much of the origins and quality of colonial and postcolonial laws is unknown, principally because most research and records concern the United States Supreme Court as well as the various state supreme courts. Consequently, we know very little about the dispensation of justice by the lower courts on a daily basis to ordinary people. And, of course, it is these lower courts and courts of first instance that are most immediately sensitive to the prevailing values of, and environmental demands upon, a populace in a new, often hostile, and demographically and economically expanding period. But there are certain procedures and anecdotes we do know about these courts that can help us understand how initial judicial systems, for example, might evolve in space communities.

In many respects, the quick availability of English, or Anglicized law to the American colonies was as unfortunate as it was godsent. Along with readily available English law and pleadings came the English-trained lawyers and judges with their professional trappings and prestige that ignored the economic, civil, and criminal realities of colonial life. The daily requirements of commercial and domestic frontier life were not the same as those in cosmopolitan London, and much of the formalistic and complex pleadings in a court action were unnecessarily time consuming and did not address the needs of a frontier society.

In fact, left to the development of their own devices for social orderliness, the colonies may well have come up with legal procedures, pleadings, and substantive law of a sensitivity, responsiveness, effectiveness, and simplicity that can be born only of a pristine society in a unique environment. But, because of immediate judicial needs, and the ready availability of familiar legal regimes, highly complex English legal procedure was plunked down almost summarily and imposed on a not altogether receptive frontier mentality.

It is unfortunate that the same tendency is developing in the exploration and, more recently, exploitation of near and outer space. Lawyers and statesmen are making—even intellectually projecting—laws and entire legal regimes for humankind's use of space.

But, as in colonial, postcolonial, or any imperialistic circumstances (economic, political, or military), the actual development of the space frontier will not be by lawyers and judges. Rather, the development and practical use (as opposed to "exploration" undertaken by governments) will be through the efforts of the merchants, business people, stockholders, multinational corporate personalities, and the ever-present military. These entities already view international space treaties and related complex domestic laws as obfuscations and wishful thinking.

Just as reforms of complex legal procedures and stuffy substantive law began in England itself in the 1830s, perhaps the neutralizing of private initiative in space development by the multitude of well-intended treaties oriented toward restraining that type of initiative, and the more inflexible domestic regulations such as those spawned under the patent and proprietary data, the Occupational Safety and Health Act, antitrust laws, and investment and banking laws, will provoke the demand for reform before all private interest is suffocated. Put another way, eras of exploding industrial activity in the past have demanded simplification of laws and reform of confounding legal pleadings so that business can be "gotten on with." This was one of the primary objectives in establishing the mercantile courts, as discussed above.

Laws directly applicable to individuals inhabiting space manufacturing facilities, manned solar power stations, and deep-space exploratory communities are apt to show a tremendous simplification of civil and criminal laws and procedural safeguards. This is a direct parallel with embryonic colonial law and the early decades of Western development, when laws were simple and responsive to the needs of the self-reliant person living in a small, self-reliant town, or existing from day to day on a hostile frontier. In this respect, procedural law was kept simple because of a traditionally deep fear of the concentration of power that accrues from the complexity and opaqueness of legal pleadings and fictions formulated by governmental institutions.

In the American West, around 1800, territories close to the unpopulated frontier edge developed their own laws. A kind of broad tutelage existed in those territories, neighboring states, and the U.S. Congress, which for the most part had veto powers over territorial laws. According to Friedman, "Trained legal talent existed on the frontier, alongside fraud and animal cunning. The level of legal sophistication in a community depended on its size, on its economic base, and on whether it was close to centers of government." Friedman quotes even more colorfully regarding the lack of orderliness in frontier development, from Joseph G. Baldwin's *The Flush Times of Alabama and Mississippi* (1957): "Many members of the bar, of standing and character, from the other States, flocked in to put their sickles into this abundant harvest" of easy frauds, unnecessary litigation, and spoils of unsophisticated politics.

At the outset, there was a great deal more orderliness in the occupation and exploitation of space, simply because there was plenty of time for Earth-sitting lawyers to contemplate the phenomena of this new frontier. But a fair guess would be that the exigencies of military requirements and the nature of commerce in a new frontier (and a physiologically alien life-support medium) will induce all sorts of quick dealings, paper corporations, frauds, or otherwise questionable activities.

The tendency of isolated communities in unique environments is to publish codified laws immediately as a means of holding out a sense of permanence and independence to other more secure communities. Traditionally, the quickest way to do this is to adopt proven laws and codes, regardless of whether they are responsive to local needs. This same tendency showed itself in the early lengthy debates among statesmen, lawyers, law teachers, and students as to whether a "Space Code" should be established in the nature and spirit of the *Code Napoleon*, or whether space law should develop hand in glove with experience, much as the law of the high seas. Although the code approach was abandoned as a concept, it has sneaked in the back door in less grandiose terms in the form of a multitude of loosely connected, disparate treaties, conventions, and regional and bilateral agreements—both civil and military.

And again, much of the civil treaty law dealing with space exploration and use represents wishful expressions of what ought to be, not what is or is likely to be. The only treaty law that attempts to reflect practical experience is that which is integral to the bilateral and regional treaties on arms control, testing, and disarmament.

Impact of Economy and Commerce on the Social and Political Structure of a Space Community

Traditionally, transportation is an important factor in the economic infrastructure of a community. Commerce is stimulated by the facile movement of goods and people. Roads, canals, shipping, and railroads were early investments of colonial, state, and federal governments. Transportation is the principal stimulus for commercial intercourse and development.

Historically, most commercial laws were initiated and applied by the colonies and the states, not by England, France, Spain, or the eventual federal government after independence. Very likely, communities of space manufacturing facilities will establish their own laws to regulate internal commerce. But with the dependence upon a multinational space transportation system, commerce between space communities and supporting Earth governments and private and public consortia will quite likely be guided with a keen eye to the needs of the latter.

Although "economy" and "commerce" beget a bottomless pit of rules and regulations, those rules become stretched, distorted, and broken by the special demands of great distances and demanding physical environments. This was true in the colonial and postcolonial frontiers. What was good in the large cities on the Eastern seaboard was simply unresponsive, hence unfit, for the needs of isolated settlements, remote trading posts, thinly scattered populations, and a variety of unique social conditions. In these instances, the commercial infrastructure and its control passed from England, to colonial officials, to postcolonial governmental officials in Washington and major population centers, and thence into private hands on the frontier. Laissez-faire, as it turned out, was much more a *practice* in the 1800s than a theory—and likely will be the practice of commercially productive space communities among themselves and with Earth.

In fact, between 1776 and 1850, most government investments in "private enterprise" to stimulate or induce private commercial investment failed predictably and miserably. Many states tried to sell off their assets and delegate taxing authority just to achieve fiscal and political stability—if not survivability. One method for space facilities to exact a certain amount of control over their commercial, perhaps even political, destinies, is to help Earth governments underwrite increasingly more efficient space transportation systems for the purpose of creating more trade markets. And this would require an early policy that demands an independent economic base for all space communities. It would require that the commercial relationships between "Spacekind" and "Earthkind" offer parity in profit return. This, in turn, demands that we continue to evaluate carefully the philosophic, as well as practical, constructs for occupying and exploiting near and deep space in the first instance.

Many socioeconomic and political structures have been suggested for space communities—some reasonably disciplined and many more resulting from unstructured wishful thinking. It has been a time of casting about for a workable meld of sociopolitical principles and new economic theory to induce the building of space communities and the stabilizing of their populations.

One concept which seems to have been overlooked and which finds its origins in the late 1700s, early 1800s, is the "Massachusetts" or "business trust." This was a sort of unincorporated association of people that conducted business in the form of an English trust where managers, or trustees, held title to the property, principal, or basic marketable resource for the use of beneficiaries. Although this "business trust" structure as the framework through which a government body carried out its responsibilities was out of place amidst the pressures of big business and was outstripped by evolving "corporate structures," it might form the basis of a small, commercially oriented society in a space habitat.

The whole area of noncommercial social law, both civil and criminal, in space communities also might take a chapter from U.S. history. Inevitably, marriage and divorce laws, as well as sanctions against violation of criminal laws, were formed by a region's practices and sentiments—despite the assertion that state and church were the keepers of social morals. This is directly reflected in the creation of common law marriages, property ownership by women (encouraged by frequent desertions into the vastness of the frontier by husbands with burdensome debts, criminal records, or simple wanderlust), and enforcement of antislavery laws. In the same vein, rather than imposing Earth-indigenous, unresponsive, and insensitive social or personal status laws on space community inhabitants, they are likely to evolve from the prevailing values of the space inhabitants themselves.

Conclusions

The social, political, and economic forms of a space community will depend principally upon one very influential and obvious factor: the source of funding. Many theories regarding that source are being talked about by the early planners, such as:

1. International banking organizations, both public and private;
2. Multi-national corporations;
3. Insurance underwriters with large amounts of capital and much experience in risk investment;
4. A strictly military–private enterprise venture sharing the common interest of protecting satellites and orbiting communities from hostile attack, and the inherent need of the military for ongoing development in tactical and logistical weaponry;
5. Periodic support of a "space rush," similar to the land rushes held by the U.S. government when it was long on debts, short on cash, and far on land; and

6. International governmental cooperative efforts.

There are as many theories on how to institutionalize space communities as there are creative jurists, statesmen, economists, and prudent investors. However, despite the hopeful attitudes of the statesmen, jurists, and others who have attempted to mold what they think the sociopolitical and economic structures of space communities ought to be, the chances are great that we are entering another era of imperialism, with all the trappings of imperialistic warfare. And for that, history has shown us that we must first establish communities not of free trade and thought but of exploitation from which wars of independence spring. The long history of Earth colonizations tells us that our lofty wishes to the contrary, particularly as they were articulated and emphasized in the Outer Space Treaty of 1967, space colonies will be established with social, commercial, and political characteristics with which we have become familiar.

About the Contributors

Harry H. Almond, Jr., is professor of international law at the National War College, Washington, and also serves as an adjunct professor of strategic studies, Georgetown University, Washington, D.C. He has received the J.D. (Harvard Law School), LL.M., and Ph.D. with distinction (London School of Economics and Political Science, London). He is a member of New York State Bar; Bar of the United States Supreme Court; and is Barrister-at-Law, of Gray's Inn, London.

William S. Bainbridge is an associate professor of sociology at Harvard University. He previously served in the Sociology Department of Washington State University. His book, *The Spaceflight Revolution* (Wiley, 1976), has been widely reviewed and is now being republished by Krieger Publishing Company.

Thomas Blau works on Capitol Hill and has been a consultant on outer space issues with U.S. government agencies. He received his Ph.D. from the University of Chicago.

B. J. Bluth, professor of sociology at California State University-Northridge, has also worked with the NASA Space Station Task Force. Her writings have appeared in such disparate publications as *Teaching Psychology* and the *Journal of Mechanical Engineering*.

Nathan C. Goldman is an assistant professor of government and the E. D. Walker Centennial Fellow of the Institute for Constructive Capitalism, the University of Texas at Austin. He holds a law degree and a doctoral degree in political science and has been a NASA Summer Fellow.

Daniel Gouré is a senior analyst with Science Applications, Inc. and has been a consultant to the U.S. Department of Defense on space policy. He has written extensively on strategic defense, Soviet military affairs, and weapons in space.

Jürgen Häusler is a Ph.D. candidate in political science, and was a research fellow in a research project comparing energy policies of France and the Federal Republic of Germany, University of Konstanz. He now teaches political science at the University of Frankfurt.

James Everett Katz, a sociologist, is on the faculty of the Lyndon B. Johnson School of Public Affairs, University of Texas, Austin. He has held fellowships at Harvard and MIT and worked for the U.S. Senate before moving to Texas. His seventh book, *Congress and National Energy Policy*, is also published by Transaction (1984).

Jean-Louis Magdelénat is the assistant director of the Institute of Air and Space Law at McGill University, Montreal. He has earned a Docteur of Laws from Aix-En-Province, an L.L.M. from McGill University, and a Lic. Droit (D.E.S.).

Hans Mark received his Ph.D. in physics from MIT and has done pioneering research on X-rays from stars. He has held a variety of academic posts in both physics and engineering and has served as Director at NASA's Ames Research Center, Secretary of the Air Force, and Deputy Administrator of NASA. In 1984 he was named Chancellor of the University of Texas system.

John Joseph Moakley, a Democrat, represents Massachusetts' Ninth District (Boston) in the United States Congress. During World War II he served in the Navy. Congressman Moakley is a subcommittee chairman of the Rules Committee, and has become a strong opponent of space militarization. He has introduced several House resolutions to ban antisatellite developments.

Joseph N. Pelton is executive assistant to the director general of the International Telecommunications Satellite Organization (INTELSAT). He has worked in the field of satellite applications since 1965, with North American Rockwell, NASA, George Washington University, the Communications Satellite Corporation, and INTELSAT.

George S. Robinson is assistant general counsel of the Smithsonian Institution, Washington, D.C. He has published extensively in the fields of space law and the sociology of space exploration. He received his Ph.D. from McGill University, Montreal, Canada.

Harrison H. Schmitt, a Republican, was a U.S. senator from New Mexico between 1977 and 1983. He received his Ph.D. in geology from Harvard University and was a NASA astronaut from 1965 to 1973. In 1972, he commanded the lunar module on Apollo 17, and performed research on the moon's surface.

Georg Simonis is in the Faculty of Social Sciences, Political Science Department, University of Konstanz, and is director of a research project on energy policies.

Marcia S. Smith is a specialist in aerospace and telecommunications systems with the Science Policy Research Division of the Congressional Research

Service, Library of Congress. She is the author of more than sixty articles and reports on such topics as space programs, communications, and nuclear energy.

David Stupple, who died in 1983, was associate professor of sociology at Eastern Michigan University. Interested in all forms of popular culture, he specialized in the study of flying saucer groups and traditions.

David Swift is professor of sociology at the University of Hawaii-Manoa and contributing editor of *Astrosearch*. His reports on space-related issues have appeared in *Astronomy, Omni, Astronautics and Aeronautics*, and other publications.

Ron Westrum is professor of sociology at Eastern Michigan University, where he is associate director of the Center for Scientific Anomalies Research. He is the principal author of *Complex Organizations: Growth, Struggle, and Change*, forthcoming from Prentice-Hall.